Gardeners'
World

The
Garden
Problem
Solver

Gardeners' World

The Garden Problem Solver

Year-Round Troubleshooting for Every Gardener

Foreword by Adam Frost

BOOKS

BBC Books, an imprint of Ebury Publishing,
20 Vauxhall Bridge Road,
London SW1V 2SA

BBC Books is part of the Penguin Random House group of companies
whose addresses can be found at global.penguinrandomhouse.com

Penguin
Random House
UK

This book is published to accompany the television series entitled

First published by BBC Books in 2023

www.penguin.co.uk

A CIP catalogue record for this book is available from the British Library

ISBN 9781785948220

Publisher: Albert DePetrillo
Editor: Phoebe Lindsley
Design: seagulls.net

BBC Gardeners' World Magazine: Tamsin Hope-Thomson, Kevin Smith

Printed and bound in Great Britain by Clays Ltd, Elcograf S.p.A.

The authorised representative in the EEA is Penguin Random House
Ireland, Morrison Chambers, 32 Nassau Street, Dublin D02 YH68

MIX
Paper from
responsible sources
FSC® C018179

CONTENTS

Foreword

BY ADAM FROST

For me, gardening is a lifelong journey. I have learned so much over the years from the people I have gardened with, the mistakes I have made (and continue to make!), and from just taking a moment in the garden: looking, observing, and of course making notes. The garden provides a rhythm to my life and a connection to the natural world that I really seem to need, whether I am sowing, taking cuttings, weeding, pruning, making compost, growing vegetables for the kitchen or just sitting.

Having your hands in the soil seems to replenish the soul. The physical and therapeutic benefits of gardening have been well established, and there's something to be gained from every little thing you do, whether it's planting a new tree or organising your shed. I know for some people gardening can seem a little daunting, especially if you're new to it. But it's all about having a go. If you get stuck in, you'll learn on the job. And for as long as you garden, you'll never stop learning. That's a wonderful thing.

My own love of gardening started early. As a young lad I could often be found following 'Tidy Nan' and Grandad around the allotment, dropping the leeks in holes, lifting potatoes, and planting out marrows. My favourite job was being allowed to open the greenhouse in the mornings. I would run down the path and open the door, in hope that one of the tomatoes had turned red enough that I could pick it. The smell was incredible, and that memory returns every time I catch the scent of a tomato to this day.

If I wasn't with 'Tidy Nan' I could be found just up the road roaming in 'Scruffy Nan' and Grandpa's garden, which was a little on the wild side. Nan used to collect and hoard, and one of her favourite things was Belfast sinks; they were all over the garden, each an individual world, and for a kid it was magical. I would lose hours there. It was 'Scruffy Nan' who first taught me to propagate; there were two aluminium greenhouses at the bottom of her garden, one full of cacti and the other coleus, the painted nettle. I would spend ages pricking out the little babies of the cacti and potting them on, then taking tip cuttings from the coleus. I earned my pocket money by growing these plants on and then setting up a table outside Nan's house and selling the cacti and coleus I had propagated to passers-by. Their garden was about exploring and being free, and there is hardly ever a day when I'm in my garden and one of them doesn't get a thought.

My professional gardening life started when I was 16 years old, working as an apprentice for the North Devon parks department. We did a little of everything: looking after Victorian parks, growing bedding plants, maintaining sports turf, caring for nature reserves, tree work and more. But looking back on that time it was the people that I remember most; every one of them was happy to give me time and share the knowledge which they had built up over the years.

At 21 I got an interview for a job with the late great Geoff Hamilton (who was around at the birth of *Gardeners' World* magazine). I remember following him around his garden just listening to him talk about his plans for the place. The man was completely ahead of his time – even over 30 years ago he was into organic gardening and peat-free compost and spoke passionately about the destruction of our countryside. I was lucky enough to get that job. I soon fell under his spell and learned so much from him and his son Nick Hamilton. I still miss him, but he is my inspiration, and looking back I can see that he set a sort of moral compass for how I garden. He was never afraid to make mistakes, and when you see that first-hand it gives you confidence. He also seemed to garden not only for the joy of it, but to aid the natural world – something that's stuck with me over the years.

I have created quite a few gardens over the years, and I always find that one of the most rewarding parts of that journey is seeing how gardening can help support the wildlife around us. Providing food and improving habitats for wildlife is a huge part of what we do as gardeners, and this seems to become more and more important every day.

It has taken me a long time to work out why I garden. In the end I think it comes down to this: for me the garden is a safe place, one where I can just be myself, slow down and be at ease, whether I'm teaching my kids where their food comes from, pruning the roses, planting a new tree or just weeding. There is a lot of talk nowadays of the value of being in the moment, and that is partly what drives me: how many moments can I enjoy throughout the year? I don't worry too much about chasing perfection. Things come and go – that's gardening.

Looking back, I have been lucky to spend time with so many wonderful gardeners, from 'Tidy Nan' and 'Scruffy Nan' to Geoff Hamilton and many others along the way. Every one of them was happy to spare their knowledge, and I inhaled it. Gardening does take time, and we will never get everything right, but the more we know the more rewarding it can be. The pages ahead are full of knowledge that has been amassed by experts over the years, with plenty of down-to-earth practical advice to help you get a little more for your patch. Everybody who has contributed to this book has their own gardening story, and I hope it can add a little bit to yours.

But first, don't forget to slow down, take a moment, and just enjoy.

Adam Frost, 2023

Lawns

INTRODUCTION

Lawn problems can seem daunting, but if your lawn starts to look a bit lacklustre it is amazing how quickly it responds to attention. A bit of spiking can work wonders to get the grass growing again, while raking out the dead grass can turn a thin, threadbare lawn back into a soft, green picnic blanket.

The biggest enemy of heavily used grass is compaction. It might not sound serious, but 'squashing shut' the tiny air spaces in the soil impedes drainage, prevents proper aeration and makes it difficult for roots to penetrate. And dry summers don't help the situation one bit. So the combination of hard soil, summer drought and winter water-logging is why a lot of lawns look so 'tired'.

Faced with grim-looking grass, people often feel the only solution is to rip the lot out and start again from scratch, even in spite of the time it takes to establish grass from seed, and the cost of using turf for 'instant' results. In fact, all that's usually needed to remedy the situation is a bit of care and attention.

'A lawn looks soothing – the colour green is visually neutral and acts as a cooling foil for colourful borders, and a pleasing contrast to paving, gravel and paths.'

Alan Titchmarsh

FIVE COMMON LAWN PROBLEMS

Compaction

The solution is to aerate the soil by pushing a garden fork about 10cm deep into the soil every 10–15cm and gently rocking back and forth on the fork handle. The metal points open up the soil, relieving the compaction, allowing the roots to breathe and encouraging the grass to regrow and fill gaps. On heavy ground, such as clay soils, brush sharp sand or some fine horticultural grit into the holes to improve drainage and stop future compaction.

Dandelions

Broad-leaved weeds like dandelions and plantains develop flat pads of leaves that shade out the grass, and by lying flat on the lawn these large leaves can escape the mower. Having picked off the flowers to stop them seeding, use a pointed trowel, daisy grubber or old knife to dig down into the soil around these weeds to remove their deep roots. If regrowth occurs, repeat to weaken the plant.

Moss

Moss is simple to deal with in the short term. Scarifying in the spring will give your lawn plenty of time to recover before summer arrives. Scarifying removes moss, dead grass and thatch from your lawn, which helps improve the health of your grass. To do this job, you can either use a spring-tine rake or use a mechanical lawn scarifier. Pull the rake through the grass so it's touching the soil. If you feel this is too much like hard work, electric or petrol scarifiers are available to buy or hire. Once you have scarified the area, over-seed it and water well until the seed takes. In the longer term, it's worth working out what caused the moss in the first place – for example, compacted soil, mowing too close, shade – and dealing with that to solve the problem once and for all.

Dips and bumps

Level dips and divots to get your lawn looking its best. Slice an X across the centre of the sunken area with a spade. Peel each triangle of turf back to the level area. Fork over the exposed soil and add some loam or topsoil from a border. Rake it flat, firm it with your feet and add more soil to bring it up to the level of the surrounding soil. Fold the peeled triangles of turf back down, patting them firmly level so there are no gaps. Deal with a bump in the same way, by removing soil rather than adding it.

Bare patches

One of the most common lawn complaints is bare patches, usually caused by wear and tear or accidental damage. You can buy turf to replace these patches, but as there are hundreds of different grass variety blends it is unlikely you will be able to match your lawn. You can solve this problem by making spare strips of grass from leftovers when re-shaping or edging the lawn, so they blend in seamlessly. Just collect up the strips and place them 5cm apart in a compost-filled seed tray, then grow them on outside or in a cold frame.

To replace the bare patch, cut out a square or rectangle around the area and, using a hand trowel, dig up the soil in the rectangle to whatever depth of soil your new turf strip is. Gently rake the space over with your hand (use a fork or rake for larger areas), and then lay the turf, cut to fit, over the patch. Firm it down so that there are no gaps and your new turf is no higher or lower than your existing lawn.

Alternatively, loosen hard bare soil by pricking over the surface with a small hand fork after removing any weeds, moss or dead grass, then sprinkle seed sparingly over the area and cover thinly with sieved topsoil or seed compost. Mark the area with canes to avoid mowing over the repair.

'Any time there is a spell of drought OR heavy rain, leave the grass to grow rather longer than usual – mow it higher by raising the blades an inch or so, and mow it less often. That's the best way to reduce stress straight away, since it relieves the roots of a big burden. Then put things right in autumn; September is the very best time.'
Alan Titchmarsh

ACTION PLAN
HOW TO KEEP YOUR LAWN HEALTHY

'When I talk to people about their grass, they usually say, "I haven't really got time to create a good lawn", or, "There's no point as the kids will ruin it anyway", but it needn't be that way. Spend just a few hours on your lawn and you'll be all set for the season ahead – it's really not difficult.' **Adam Frost**

To water or not?

It's important to water young lawns to help the grass put down roots, but don't overwater as this can cause them to root shallowly and establish poorly. If possible, use rainwater from a water butt or grey water from your bath or washing up bowl. Sprinklers do keep lawns hydrated but they use a lot of mains water and are not permitted during a hosepipe ban. Don't worry if an established lawn goes brown, it's not dead. It will green up again when the rains come. It's a good idea to raise your mower blades in summer – longer grass goes brown less quickly, because the plants are less stressed from being cut back.

If you don't want to water your lawn too often, choose a tough fescue, which is a variety of grass that spreads by rhizomes, filling bare patches easily. Its deep roots allow it to stay green and survive drought for longer.

Tackle weeds

The worst weeds are plantains and dandelions, as they have wide, flat leaves that can smother large areas of lawn. Their size, though, is their downfall, as they are easy to trowel out, roots and all. Yellow medick, buttercups and clover can also spread quickly through a lawn, so raking before you mow can help to lift them up into the mower blades, thereby weakening and killing them off over time. Regular mowing and feeding to get a healthy lawn is a better and more environmentally friendly weed preventative – particularly if you have pets and young children – than using weedkillers. Chemicals are expensive and ultimately don't address the reason that the weeds are there, which is poor grass health.

Feed for lawn health

In March, when the grass starts to grow, buy a balanced lawn fertiliser. Don't overdo it, and avoid the borders, otherwise your garden will look like you have run amok with a flame-thrower. If you have a large lawn, invest in a wheeled lawn feeder for a fast and accurate job. Try to apply fertiliser to grass when rain is forecast, so that it gets washed down to the roots and to stop it burning the leaf blades. If it doesn't rain, water the fertiliser in well. Your grass should look greener within a week.

Feed again in midsummer, but don't feed newly sown grass or recently laid turf as it will scorch and turn brown. A good organic alternative to chemical lawn fertilisers is liquid seaweed, which you apply once a month from spring to late summer. This helps prevent problems and gives a boost to grass growing in difficult areas, such as under trees. Seaweed is a stimulant that increases the vigour of plants when conditions are less than ideal, thereby warding off weeds and moss that tend to invade poorly growing grass.

Use a fork to aerate your lawn

Without air at the roots, grass becomes very shallow-rooted during the winter, making it more prone to drought when the water table recedes in summer. The grass will grow less vigorously, which allows weeds to get a foothold. The key to minimising these problems is aeration – achieved by spiking the lawn thoroughly to get air to the roots. A fork with narrow tines (such as a border fork) is ideal, and you need to push in the tines to at least half their depth, and wiggle slightly before you pull them out. Aim to spike in rows, at 10–15cm intervals, and finish by brushing sharp sand (not building sand) into the holes to stop them closing up.

Improve drainage

The surface of a well-used lawn can become very compacted. This is the time to relieve the pressure by using a hollow-tined aerator to remove cores of soil, which opens up the ground. Sweep up the cores and brush sharp grit or sand into the holes, to create drainage channels. Powered scarifiers can also help, but their blades usually only just penetrate the soil surface, while hollow-tined aerators go far deeper.

Top-dress your lawn in autumn

Lawns are often expected to look great year after year with the minimum of nutrition, so give your whole lawn a top-dressing of gritty, loam-based compost, or buy bags of lawn dressing for the job. Cut the lawn first, then sprinkle on the lawn dressing. Brush it down onto the soil surface using a soft broom. Rain will also wash it down, or you can use a hose end sprinkler, so the nutrients reach the roots.

Scarify to improve grass health

After mowing with a grass box to catch the clippings, rake the lawn vigorously with a springy, long-tined lawn rake or electric lawn raker to remove moss, creeping stems and trodden-in decaying material. This will allow air to circulate around the leaf blades to keep grass in good health. It also cuts through the stems of creeping grasses, encouraging them to clump and thicken up.

PROJECT – How to repair damaged lawn edges

It's easy to repair damaged lawn edges. You'll need some grass seed and loam-based compost to fill in a few gaps, but the seed should quickly germinate and the lawn should be perfectly green in a few weeks. If the edge has dropped, raise it up with some soil before laying the turves.

1. Look for places where the lawn edge has been scraped by the mower, died back under plants flopping over or has been trodden and worn down over the years.
2. Use a spade or half-moon cutting tool to cut out a rectangular section of turf around the damage, pressing down to the base of the grass roots.
3. Lift out the turf section with a spade at a low angle or an angled turf iron. Even out the level of the soil in the hole and trim the depth of the turf to make it even.
4. Turn the turf around and line up the good face with the edge, then add compost and grass seed to the gap left in the lawn by the damaged edge.

SPOTLIGHT ON
MOWING

Mowing is the key to a healthy lawn. Each cut encourages the grass to grow more thickly, creating a luxuriant look while blocking out weeds and making grass more hard-wearing.

~ **Reduce lawn height** by only one-third each time you mow – any more would decrease the grass's health and vigour.

~ **Cut fortnightly** in early spring when grass growth is slow, but make sure you increase this to once or even twice a week when growth rates peak in late spring. If you do it less, you'll break the one-third rule, as you'll have to cut off more than one-third of the grass's height to keep the lawn looking neat.

~ **Adjust the mower's cutting height** to suit the weather. When it's hot and dry, raise the mower blade to let the grass grow longer. The extra moisture held in longer foliage helps to keep grass green and shades the soil below.

~ **Mow on dry days.** If you cut when the ground is sodden, the mower will smear and rut the soil, while wet grass clippings will clump and smother the lawn below.

~ **Get the height right.** Most lawn grass is best kept at 3–4cm tall. In patches that get more wear, leave it a little longer – say 4–5cm – and in shade even longer still at 7–8cm.

~ **Get your mower serviced** at the start of the growing season to ensure a clean, quality cut. Chipped or blunt blades bruise the grass and cause straw-coloured die-back at the tips.

3 top tips

- **Leave the lawn longer** in summer, around 5–10cm. This will make it far less prone to drought, and shady lawns will also benefit as it will make them less prone to moss and bare patches.

- **Look after your mower.** Keep it well maintained and the blades sharp. Cutting a lawn with blunt blades bruises the grass and makes fungal infection more likely.

- **If you walk over your lawn** regularly, set stepping stones into it or sprinkle 1cm of horticultural grit over the area that is most used to reduce compaction of the soil.

LAWNCARE CALENDAR

Spring

~ Grass grows whenever temperatures rise above 7°C. Mow the grass lightly in a mild March. Cut grass fortnightly while growth is still slow

~ Grass will be growing fast by April – start mowing once a week

~ Lay new turf

~ Sow new lawns

~ Aerate your lawn (see page 9) and rake up moss

~ Tackle any weeds

Summer

~ Mow the grass weekly, except in hot dry weather

~ Raise the height of the cut in hot, dry weather and mow less frequently

~ Don't let turf dry out before it has a chance to root

Autumn

~ Collect any fallen leaves. If they are left on the lawn they can cause grass to turn yellow. Rake them up or use your mower to chop them up and then add them to the compost heap

~ Sow new lawns

~ Pull out any weeds

~ Spike and brush in sand

~ Rake out thatch

~ Tidy lawn edges

~ After a rainy spell when the ground is moist, sprinkle autumn lawn feed or organic fertiliser all over the lawn, following manufacturer's instructions. This will toughen up the grass and develop strong roots without causing soft growth that needs extra mowing

~ As conditions turn cool through autumn and lawns grow more slowly, raise the cutting height of your mower to leave grass longer. A cutting height of around 4cm is ideal for most lawns, while for rougher areas 5–6cm is best

Winter

~ Stop mowing, although in mild areas you may want to mow until the weather gets colder

~ Try to avoid walking on the grass, especially during cold weather

~ Get your lawnmower serviced if necessary

~ Clean lawn tools

Q&A
COMMON QUESTIONS ABOUT LAWNS

Should I leave grass clippings on my lawn?

In the summer, it's usually too dry for the mowings to decompose, so they are best removed. However, in spring and autumn, if you cut your lawn once or twice a week, the moisture in the soil will help the clippings break down quickly, adding valuable nutrients to the grass, so you can make every other cut without a grass-collecting box. The key to success when mowing without a collecting box is to cut little and often so that the lawn isn't swamped with clippings.

How do I get rid of lawn fairy rings?

The root-like mycelium of this fungus is dense and water-repellent, so prevents moisture reaching the roots of the grass, which then dies of drought. The toadstools appear in rings on the lawn and these get larger year by year as the fungus spreads. Around the edge of the ring and inside it, rings of lush, dark green grass develop, but the area between the inner and outer ring is dead and brown. Fairy rings only attack turf grasses, but everything from a well-tended lawn to rough grassland may host this fungus.

Brush off and collect the toadstools as soon as they appear, ideally before the caps open, as this will stop them spreading their spores. Only mow infested areas once all the toadstools have been binned or burned. Regularly make drainage holes in the infested area with a fork to allow water to penetrate.

If you're really determined to get rid of the fungus, then remove the infested area of grass and the soil beneath it and to either side of it, to a depth of at least 30cm. Dispose of the excavated soil well away from the garden.

How can I revive a brown lawn after hot weather?

Brown grass doesn't mean the lawn is dead. With a little help it will recover. Scarify in autumn, then spike the lawn with a garden fork. Brush an equal mix of compost and gritty sand into the holes to improve the soil's moisture-holding capacity and drainage. This combined with autumn rain will help green up your lawn. Avoid overfeeding your lawn as this can lead to too much soft growth that is more prone to drying out in hot weather.

What can we do about clover on the lawn?

Clover is good for pollinators, so consider leaving it. If you do want to remove it, lift it up with a spring-tine rake. Feed your lawn in spring and autumn and scarify in autumn.

What can I plant as a wildlife-friendly alternative to lawn?

In a sunny spot on free-draining soil, try carpeting plants like fragrant non-flowering chamomile, *Chamaemelum nobile* 'Treneague', and grey-green *Acaena novae-zelandiae*, which turns bronze-purple in autumn. Thyme flowers are pollinator magnets, so add low-growing *Thymus serpyllum* 'Russetings'. You could also plant succulents like sempervivums and *Sedum spurium* 'Green Mantle' with its starry yellow flowers. If you have moister soil, try a mix of pink-flowered *Geranium sanguineum* 'Vision Violet', darker pink *Phlox subulata* 'Purple Beauty' and thrift (*Armeria maritima*) with pink or red pompoms. *Leptinella potentillina* has tiny ferny leaves that turn from green to pewter.

My lawn was covered in ant hills last year. What can I do ?

Ants love light soil. Their hills will ruin the finish of a smooth, green sward, and one advantage of settling for a lawn of grasses and wildflowers is that lumps and bumps tend not to show.

When ant hills are flattened by mowing, the smeared soil is perfect for the germination of weed seeds, so brush them off first with a broom to spread out the soil. Do this when the soil is dry; or, if it's a little moist, add some sand before brushing.

There is also a biological control. The nematode *Steinernema feltiae*, which is watered onto the lawn in solution, is not tolerated by ants and drives them away. The effects of this treatment usually last for six weeks.

I want to reduce the size of my lawn – what can I do with the turf?

Pile up the pieces of turf in a stack, grass side down, in a shady spot, such as behind your shed, and leave for 12 to 18 months. The grass will rot away, leaving topsoil that you can mix with potting compost for long-term container plants, or simply re-use in borders wherever soil levels are low.

What can I do to make my lawn more level?

The easiest way to overcome dips and bumps in lawns of any age is by regularly applying a loam top-dressing to the surface – in autumn, and again in the spring. This material will gradually build up in the hollows and the grass will grow up into it to make the lawn more level. All but the largest hollows can be evened out in this way over the course of two or three years. Use a shovel or spade to distribute the loam evenly across the surface of the lawn and then spread it out with a stiff besom or garden broom. Avoid walking on the treated area for a week or two to allow the grass to grow through the loam and prevent it smearing across the surface of the lawn.

How does thatch form?

The fibrous material that builds up in an area of lawn surface is known as thatch. It forms from a mixture of the dead bases of grass stems and leaves, as well as any clippings. During warm, moist weather, soil bacteria and fungi will work on this layer to break it down, returning the organic matter to the soil. In dry conditions – most likely in summer – this decomposition stops and the thatch becomes mummified in a water-repellent layer. It's a good idea to rake out this dead material, allowing air to circulate around the grass plants which, in turn, will help prevent fungal infections and die back.

When is the best time to sow a new lawn?

September is the ideal month to make new lawns from seed. The soil is still warm after the summer, and autumn rains can usually be relied upon to keep it moist, providing the perfect conditions for germination. This means the young grass plants should be large enough and have a sufficiently deep root system to survive the winter. In fact, once they are initially established, the colder weather will suppress vigorous growth and promote 'tillering' – the clumping up and production of sideshoots – of each grass plant. In addition, the new lawn will be able to establish at a time of year when you are less likely to be using the garden.

Compost

INTRODUCTION

Homemade compost is useful, its production involves little financial outlay, it helps to conserve our valuable peatlands by not depleting them and it will save you the expense of acquiring other forms of soil enrichment. Where once you bought the stuff in plastic sacks, with a little planning and effort you can become self-sufficient. Your soil will benefit, your pocket will benefit and you'll feel supremely virtuous!

When used as a mulch, homemade compost helps to suppress weeds, retain soil moisture in dry weather and protect soil from the damaging effects of heavy rain. Once incorporated into the soil (there is no need to dig it in, the worms will do that), it improves soil structure, helps to retain water and, as it breaks down, releases essential plant nutrients, such as nitrogen. The product you make has huge value – plenty of organic matter means lots of life below ground, supporting the whole garden ecosystem. It's true that creating compost doesn't always go to plan, but it's easy to solve any problems that come up if you follow a few simple rules.

'I love compost. I love the alchemy of taking all the waste from a garden and household and turning it into a sweet-smelling mix that does more to improve your garden than anything you could possibly buy or bring in from outside.'

Monty Don

FIVE COMMON COMPOSTING PROBLEMS

Smelly wet heap

The secret to successful composting is getting the right mix of ingredients. Nearly all compost problems are the result of a wrong balance, many of them stemming from too much moisture. Most kitchen and garden waste, especially piles of lawn clippings, tends to be wet and low on structure, which can lead to a heap that's wet, airless and smelly. The solution is to add scrunched-up paper or card, such as egg boxes.

Slow compost

Turning your heap will speed up decomposition, but not a lot, and ultimately the rate at which you make compost is determined by the rate at which your garden and kitchen supply the raw materials. The other way to speed things up is to use two bins: fill up one bin completely, as a full bin will heat up and decompose faster, then use a second bin for your waste while the other is maturing. You could also try adding nitrogen-rich materials such as fresh horse manure, comfrey and nettle leaves.

Flies

The larvae of many flies feed on decaying organic matter, so flies are a normal part of the composting process, but you can discourage fruit flies by burying fruit waste beneath other material such as grass clippings.

What to put in

- Green waste from the veg patch
- Bedding plants, at the end of the season
- Soft green prunings
- Faded flowerheads
- Crushed eggshells
- Torn-up newspaper
- Annual weeds
- Grass clippings (but mix them in!)
- Autumn leaves in small quantities
- Vegetable kitchen waste, such as carrot tops and potato peelings

What to leave out

- Diseased plant material
- Cooked food (including bread and mashed potato)
- Thick-rooted perennial weeds
- Annual weeds that have run to seed
- Thick, woody stems (unless shredded)
- Large amounts of autumn leaves (make leafmould from them separately)

Weeds

The high surface area:volume ratio of a typical small domestic compost heap means it loses heat too quickly and won't get hot enough to kill diseases or weed seeds, so don't add them to the heap in the first place. Nor will it kill the roots and rhizomes of perennial weeds, but these can be composted as long as you kill them first, leaving them to dry out before adding them.

Dry heap

A heap won't work if it's too dry. Perhaps it contains too much dry, twiggy stuff, although that is fine to use as long as it's chopped up first. Or maybe your bin has too much ventilation or lacks a lid. The ideal compost heap is about 50–60% water – if you can just squeeze a few drops of moisture out of a handful from the centre of your heap, that's perfect. If your heap seems too dry, water it.

ACTION PLAN
HOW TO MAKE GOOD COMPOST

~ **Mix the contents evenly** as you add them, with no big concentration of any one material in any one place – lawn mowings, for example, turn slimy or white and fungus-ridden if not mixed with larger-leaved waste. Too much carbon (woody stems for example) and the composting process will be too slow, but too much nitrogen (grass cuttings, leaves and so on) and you'll get an evil-smelling sludge. For every load of green material, mix in the same volume of dry material such as shredded stems, straw or cardboard. Aim for a 50:50 blend of carbon (generally drier, brown material) and nitrogen (green waste). Chop up or shred woody stems, so they decompose faster. You can buy or hire a shredder, and if you're a confirmed composter you'll find it wonderfully rewarding.

~ **Separate ingredients.** Remove leaves from other compost ingredients and use for leafmould.

~ **Turn the heap.** If you have the time, turn it as often as possible. Once every 10 days is ideal.

~ **Keep the contents moist.** In spells of hot dry weather, give your bin a good watering. In order to continue rotting down, moisture is essential. If the contents dry out, the composting process will slow down or stop altogether. Protect it against rain to prevent it getting too wet.

~ **Firm it down.** From time to time, either tamp down the bin's contents with the back of a fork or rake, or climb in and tread it down. Air pockets allow it to dry out in patches and rotting will be uneven.

~ **Cover the top** with a square of old carpet or sacking to keep the moisture in and help to retain the heat.

Compost facts

- **A really big** heap loses heat more slowly, and may stay around 50°C or hotter for a whole year.

- **Most tree leaves** contain too much lignin to break down quickly, which is why they're better left in a heap of their own, where fungi will slowly turn them into leafmould.

- **The pH** of most compost is around neutral (pH7) or even higher. To make acid compost you need to start with really acidic ingredients, such as bracken or pine needles.

- **Composting** without enough air produces compounds such as ammonia and hydrogen sulphide, which is why a soggy heap doesn't smell very nice.

- **Compost heaps**, like everything else in biology, work faster when they're warmer, so don't expect your heap to do much in the winter.

PROJECT – How to make leafmould

With a little extra effort you can make specialist composts, such as leafmould. Fallen autumn leaves are most useful when composted separately for this. It's slow to break down but the final result can save you money if you use it in your potting compost. But don't collect up every leaf – a few piles of leaves in odd corners provide welcome shelter for wildlife.

1. Pile leaves into post and wire-netting containers or empty potting-compost bags, sprinkling with water if they're dry. In 18 months to two years, you'll have dark, crumbly leafmould to use.
2. Tread batches of leaves down as you bag or bin them, so they take up less space.
3. Add worms to leafmould bags and bins. They will help the process of breaking down the leaves, quickly turning them into crumbly compost.
4. Then leave the heap alone. Once collected and compressed, unlike compost, leafmould doesn't need turning. Leafmould that you've collected in a wire bin should be watered if it's dry. It'll break down and decompose better if it's moist.

SPOTLIGHT ON
WORMERIES

These small-space composting powerhouses offer something for every garden. In most cases a wormery is best used along with traditional composters to process food waste only, while in tiny plots it is a neat, clean and space-saving way to recycle garden waste.

Worms work best between 10°C and 30°C, so wormeries will work slowly in winter. It takes a few weeks for worms to establish and work at full speed, so be patient.

Why use a wormery

~ Composts food waste that won't go in a standard compost bin
~ Self-contained and clean, and can stand on paving, handy for the kitchen. Could be used indoors
~ Size is ideal for small spaces
~ Faster than traditional composting methods
~ Reduces waste going to landfill
~ Provides extremely rich compost and liquid feed

Drain liquid from new waste before adding, and drain off liquid regularly to reduce condensation and prevent the wormery getting soggy and smelly. If the model lets in rain, cover loosely.

Bury fresh waste or cover with damp newspaper or brown card to avoid getting lots of tiny flies. These fruit flies, although harmless, can be annoying.

Chop up waste before adding, to avoid mould. Damp conditions are essential for worms as they breathe through their skins, so they actually like mouldy food. However, lots of uneaten waste may be a sign that you're adding too much at one go – little and often is best.

In winter move your wormery into an outhouse or porch. But outside is fine with care; cover with insulating material, put thick layers of card and paper inside the top, and drain often.

'It's such a waste not to use your plant waste – far from being useless and unwanted, your garden leftovers can become a positive benefit to your soil. That's what composting is: processing garden waste into something of value, not only to your patch of earth, but also in terms of caring for the environment.'
Alan Titchmarsh

3 top tips

- **Add insulation** – worms and other organisms will be more active if your bin is warm in winter, 'cooking' your compost quicker. You can buy insulated bins, but they are pricey. Instead, cover your bin with flattened cardboard boxes, old carpet or large polythene sacks filled with broken-up polystyrene or straw. Stand bins in a sunny spot during winter.

- **Dry out perennial weeds before adding to a compost heap** – tie up perennial weed roots in a plastic sack for a few months before adding to your compost.

- **Use compost to enrich soil** – in the veg patch or allotment, enrich your soil with homemade compost or use it as a mulch on your veg patch, flowerbeds and borders. Coupled with a sprinkling of general organic fertiliser, such as blood, fish and bonemeal, it's a matchless form of soil enrichment that improves soil structure, texture, micro-bacterial activity and overall fertility.

COMPOST CALENDAR

Spring

~ This is a good time to start a compost bin or heap, when the weather is warming up

~ Add a mix of material regularly to fill your bin

Summer

~ Check your compost bin for moisture, and water it if it appears to be too dry

~ Turn compost if you want to speed up decomposition

~ Firm down contents to get rid of any air

Autumn

~ Turn compost if you want to speed up decomposition

~ Keep adding a good mix of brown and green materials

Winter

~ Compost should be ready after six months to a year, so there may be some to use in the garden by now

~ Insulate any bins that are not being emptied

~ Composting will slow down over winter

Q&A
COMMON QUESTIONS
ABOUT COMPOSTING

How long will it take to get compost?

Green waste rots down faster in warmer weather and slows down in winter. If you bank on a yearly turn around, you're unlikely to be disappointed and the time spent tending your developing compost will be minimal. If you have plenty of time on your hands and opt for the hot composting method of filling a bin with a good mixture of material, then turning it once a week, you can produce compost in as little as three to six months, but you will need to factor in the time and effort involved.

What's the best container?

Even in a small garden, it's a good idea to choose the biggest bin you can, as the smaller the bin, the more difficult it is to make compost efficiently. Ideally it would be around one metre square and high. Two bins are even better, so you can be filling one while using the rotted contents of the other, once your system is up and running. Make the bins from wooden posts and wire netting, or with slatted wooden sides that are removable, for easy access. Although not as environmentally friendly, you could buy a second-hand plastic compost bin or get one free through your local council.

What is hot composting?

There are two main composting methods – hot and cold. Hot composting involves filling the bin to the brim with a mix of organic matter, then turning it regularly so that rotting is rapid and heat builds up. Cold composting takes longer, but doesn't need turning. Just fill the bin gradually when organic waste becomes available. As this decomposes, it heats up but not as rapidly as with the hot system. It will still rot down well over the course of six months to a year, and you can also turn the compost if you want to speed things up.

What can I do about bindweed in my compost heap?

If you leave your heap the bindweed will just carry on, pilfering all the goodness and spreading itself further. If your materials are only partly composted, you could re-mix your heap, adding extra manure or other heat-generating ingredients such as fresh grass clippings. This way, the heat given off can cook the bits of root in the mix.

If your compost is already finished, you could sieve out the roots. This is much easier if the compost is dry, so help water to evaporate from the heap by keeping it protected from rain but open to the wind. Sieving also helps to remove any bits of rubbish, or any large, half-decomposed pieces.

Our homemade compost is full of unwanted seedlings. What should we do?

Seedlings are very likely to pop up in homemade compost, as you have found – both weeds and other surprises, such as tomato plants. Weeds that are most likely to appear in compost either have pernicious underground parts (e.g. docks and nettles) or seeds. First, always avoid putting weeds with chunky underground parts of any sort into compost bins and similarly avoid weeds that are bearing seeds or are in flower. If necessary, cut off the offending parts and rot them down in water, then compost just the 'safe' bits.

Can you use homemade compost for sowing seeds?

Homemade compost doesn't tend to carry worrying pathogens. But as small composting piles and bins do not reach the temperature of industrial scale composting, they are unlikely to kill fungal spores. This is not a worry for most established plants, as many natural fungi help to break things down, which is why they are in your compost to start with. However, due to the delicate nature of seedlings, it would be wise not to sow seeds, or prick out seedlings, into homemade compost. For seed sowing, it's best to buy compost.

Containers

INTRODUCTION

Containers are an easy way to add colour and interest to your garden. Even one container can make a striking focal point for entrances, patios and balconies. They are easy to plant up and can be a good way to trial different planting styles and colour schemes. Summer containers can provide non-stop colour right through to the first frosts if you choose suitable plants and look after them well. Whether you've got a tiny balcony or a large plot, every space can be improved by a few colourful pots. To keep your displays looking their best it's important to keep on top of maintenance and any problems that crop up. Here are some of the most common problems you'll come across.

FIVE COMMON PROBLEMS WITH CONTAINERS

Containers drying out

For a start, choose a few large containers rather than lots of small ones. The greater the quantity of compost, the longer it will take to dry out. You can add water-retaining granules to the compost, but it's still essential to be diligent about watering. This is best done early in the morning or early in the evening, rather than in the heat of the day when much of the moisture will evaporate rapidly. In hot weather, you'll need to water daily.

Hanging baskets go over too early

For long-lasting, flower-filled hanging baskets, you need to water and feed them generously, as there is so little compost. They are often hung in sunny, exposed sites too, which dries them out even more rapidly. Water them at least once on bright days (morning or evening), twice on very hot days. Give them a dilute liquid feed once a week too.

Pot-bound plants

Plants are pot-bound when their roots have totally filled the container, so that hardly any compost is visible when the plant is knocked out of its pot. Many pot plants continue to flower when their roots have filled the container. In such circumstances, flowering is a defence mechanism: the plant has exhausted its nutrient supply, and feels a need to produce flowers to set seed and ensure the survival of the species. But after this, when the plant becomes extremely pot-bound and short of nutrients, it will start to lose vigour and even flowering will be beyond its capabilities.

You have two choices: you can plant them out in the garden, having teased out the roots first, or you can give them a bigger container and additional fresh compost. In the latter case, break away some of the surface crust from the old root-ball and tease out the roots from the base, settling it into its new container (ideally a few centimetres larger all round) with new compost. Water well.

Vine weevil

This pest is the bane of all gardeners – and plants in containers are especially vulnerable. The adults make notch-shaped holes in the leaves, but it is the grubs that do the real damage by eating the roots. Often it is only when the top growth is severed from the roots at the 'crown' of the plant (where roots meet shoots) that the problem is first noticed.

Make sure you buy fresh, sterilised compost at potting time, and water the compost with a biological control such as Nemasys (a minute eelworm that attacks the vine weevil) to reduce the likelihood of attack. Picking off the adult weevils (small beetles with a snout like an elephant's) is for the infinitely patient.

Slugs and snails

Defend against slugs and snails by top-dressing your pots with sharp gravel or crushed shells and by gluing copper tape around the rim of the pot. Make sure there are no leaves overhanging the tape, which slugs and snails could use as a bridge into the pot. Raising pots up on feet also helps reduce attacks.

ACTION PLAN
FOR CONTAINERS

Feed in summer

Plants in containers will bloom better and for longer if you feed them in summer, as the nutrients in the compost soon get used up or washed away. So about a month after you plant up your containers, start feeding weekly with a dilute liquid tomato fertiliser to keep the plants flowering happily.

Prepare for a holiday

If you don't have anyone to water for you while you're away, move baskets and tubs into the shade and sink their bases into moist soil, then water well before leaving. You can also snip off all the open flowers, plus well-developed buds, so that annuals are just producing a new flush of flowers when you return, instead of running to seed.

~ **Set up a watering system** with a timer to keep at-risk plants, such as greenhouse tomatoes, well watered. It can run off either an outdoor tap or a rainwater butt that can be topped up by hose or via guttering from a nearby shed or greenhouse roof.

~ **Add gel crystals** You can add water-retaining gel crystals to containers you've already planted up. Make a series of holes a few inches deep between plants with a pencil and trickle small amounts of crystals or made-up gel into the compost. This will hold moisture between waterings, releasing it slowly as plants dry out, but it also 'locks up' excess water after heavy rain.

Use less, recycle more

Sink a bottomless plastic soft drinks bottle into the soil by thirsty plants to 'funnel' water straight to the roots. You can also put a bucket by the back door to recycle veg-washing water. This saves money if you're on a water meter.

Deter pests

It's much easier to protect plants against pests when they're growing in a pot rather than in the ground, as you have a smaller area to deal with. Keep pests such as cabbage white butterflies and birds away from crops in pots with fine-gauge netting or fleece. Make a frame for the netting from canes, so that it's not touching the plants. Protect compost by placing chicken wire or spiky holly leaves over it to stop mice and squirrels digging up seeds and bulbs.

Protect plants over winter

Plants in pots are vulnerable to cold, wet weather, so if you want your display to continue beyond the summer you will need to help it. If you can, move pots to a warmer, more sheltered spot away from frost pockets, before the cold weather strikes – against a warm house wall is good. If pots are too big to move, or you have too many, protect them in position.

~ **Wrap pots in bubble plastic.** The roots of container plants are more likely to freeze in cold weather, so insulate pots, wrapping them in bubble plastic or hessian sacking stuffed with straw. Smaller pots can be sunk into the ground.

~ **Mulch your pots.** Cover compost with a thick layer of organic matter to keep it frost-free.

~ **Tuck plants up in fleece.** Protect less hardy plants from freezing winds and cold by wrapping them securely in horticultural fleece.

~ **Shelter against the rain.** In very wet spells, cover pots to stop the plants becoming waterlogged. To improve the drainage, remove any saucers from under the pots and raise containers up on feet.

'I have a small town garden and from the outset I have loved displaying containers on my patio to celebrate seasonal planting and experiment with new ideas. These "mini borders" can be tailored to suit any position or soil type ... you get to make the rules!' **Arit Anderson**

PROJECT – Plant a rose in a container

Roses are among the nation's favourite flowers and, no matter where you garden, you can grow them. Most roses are perfectly happy in pots, provided you choose the right one and look after it well – so even if you only have a balcony or tiny patio it's easy to make room for a rose. There are many different types of roses, from sprawling climbers and ramblers, to shrub, hybrid and groundcover types, right down to the tiniest patio varieties. To make sure your rose is healthy and covered in flowers, it's essential to choose the right type and the right size of pot.

- **Tiny treasures** The obvious choice of rose for container growing is the patio or miniature types, which have been specifically bred for growing in pots and have shallower roots and compact growth. Minimum pot size: 30cm x 30cm
- **Mid-size marvels** Groundcover roses and shorter climbing roses, which have been bred for growing on a patio, are bigger than the miniature types, but still compact enough to be happy in a container. Minimum pot size: 45cm x 45cm
- **Large and lovely** If you want to grow bush, shrub or climbing and rambling roses, you will need a large container, as these are vigorous plants. This is a much bigger investment in terms of pot and compost, but it's vital if you want your rose to thrive for years to come. Minimum pot size: 60cm x 60cm

1. Choose a pot large enough for the type of rose you want to grow (see page 45), with plenty of drainage holes in the base. Fill it up to two-thirds with a good-quality loam-based compost, such as John Innes No.3, mixed with a couple of handfuls of garden compost or well-rotted manure.

2. Sprinkle mycorrhizal fungi on the roots. Place a bare-root plant on the surface and spread the roots out. With a container-grown rose, remove it from its pot and plant at the same depth it was originally.

3. Fill in around the rose with compost, making sure the bare-root plant is planted at the same depth as the soil mark on the stem. Press the compost firmly down around the roots and top up with compost if the level is too low.

4. Water in well and then mulch with a 5cm layer of gravel or crushed shells to help retain moisture. Place your rose in a bright spot that gets shade for some of the day so it won't dry out too quickly and give it space so that air can circulate.

SPOTLIGHT ON
HANGING BASKETS

Hanging baskets offer an attractive way of displaying plants at a comfortable level.

~ **Choose the right compost**. Buy some that is specifically designed for pots and containers (check the label) or beef up regular multi-purpose compost with water-retaining gel and slow-release fertiliser.

~ **Use a liner that's designed for a basket bigger than yours**. You can then be sure that compost and water will stay in the basket rather than leaking out after watering or as plants grow through the season.

Place a large piece of plastic inside your liner before you fill it with compost and then cut it to fit just inside the liner. Hanging baskets hold a lot of plants in very little soil and they can dry out very quickly. This extra plastic liner will help the compost to stay moist.

~ **One way to water effectively** is to place an upended plastic bottle with the bottom cut off in the centre of your basket, but leave the rim poking out of the compost. Water directly into it to help water go straight to the plants' roots and stop it overflowing and pouring over the edge of your basket.

~ **Water every day** without fail, preferably in the evenings, and never let the compost dry out. In very hot weather you may have to water both morning and night. If you go on holiday your baskets will still need watering, so arrange for family or neighbours to take over watering duties for you.

~ **Feed baskets** with a balanced liquid fertiliser about six weeks after planting. This will boost the slow-release granules and ensure plants continue to produce flowers.

~ **Deadhead spent blooms** on a daily basis to stop plants going to seed and to encourage further flowering. Deadheading regularly will stop the task becoming a time-consuming chore and you're much less likely to miss the smaller ones.

3 top tips

- **Use peat-free,** loam-based composts for plants in long-term containers. They ensure the pot won't dry out too quickly and will hold nutrients well.
- **Raise pots up** onto bricks or pot feet so that excess water can drain away easily.
- **Plants in a** small pot quickly run out of water and food, so then produce fewer leaves and flowers or poorer crops. Wherever possible, trade up to a larger pot to help keep plants healthy. When buying a new container, bear in mind that plants are usually sold in the smallest pot they can possibly fit into, to save on cost and space. A good rule of thumb is to find a pot that you like the look of and then go one size larger.

'Our two Prunus incisa 'Kojo-no-mai' have always lived in pots – so small are they that they would be swamped in a border. In a very simple arrangement, it takes central position accompanied by Cyclamen coum.

'There is plenty to look at and appreciate in the garden, but to have on the terrace, plainly visible from the kitchen window, a pot like this – full of life and promise – is a tonic on grey days.'

Carol Klein

CONTAINER CALENDAR

Spring

~ This is a good time to plant new containers

~ Remember not to pot up tender plants until after the last frosts

~ Sow annuals for summer flowers

~ Plant summer-flowering bulbs

Summer

~ Set up a watering system if you are going away

~ Feed plants that need it throughout the growing season

~ Water regularly, especially during hot weather

~ Hanging baskets and small containers will need watering every day

~ Check for pests

~ Deadhead plants such as bedding and roses to keep them flowering

Autumn

~ Remove summer bedding from pots

~ Stop feeding plants

~ Plant spring-flowering bulbs in pots

Winter

~ Move tender plants under cover

~ Use bubblewrap to insulate tender plants in pots that can't go under cover

~ Raise pots on pot feet to help with drainage

Q&A
COMMON QUESTIONS ABOUT CONTAINERS

Which foolproof plants are best for summer containers?

Ivy-leaved pelargoniums take some beating – they will cascade over the sides of a container and flower for months on end. The ordinary 'geraniums' (strictly speaking zonal pelargoniums) are a great choice too. They love bright light and can cope with drying out every now and then. Cosmos are good too, as are bright French marigolds.

What can survive in pots that you forget to water?

Not much! Left dry, most plants will soon turn up their toes, and you cannot rely on rain to soak the compost as the foliage often acts like an umbrella. Summer bedding in particular – especially in hanging baskets – is unlikely to fully recover from drying out. The plants will have wilted, the root hairs that suck up water will have died and the plant will need to send out fresh ones before it has a hope of recovering.

If you know you are going to be absent from time to time, and your container plants are going to have to cope, then cacti and succulents such as aeoniums, sedums and sempervivums are a good bet. As are rosemary and lavender. California poppy is a bright annual that is drought tolerant, and New Zealand flax (phormium) and yucca are dramatic plants that are quite forgiving of dry conditions.

How can I keep plants in pots looking healthy?

You must remember that you are totally in charge of their well-being. They may be outside and exposed to the elements, but they are vulnerable in two ways because they have a limited amount of compost in which to sink their roots. This restricts both the food and water available to them.

Regular watering (whenever the compost feels dry) and feeding once every week or so with dilute liquid feed during the growing season are necessary to keep them happy. That said, overwatering (keeping the compost soggy) will result in plants that wilt, never to recover. This is exacerbated if the drainage holes at the base of the pot become blocked. In short, good drainage, sufficient water and regular feeding is the best recipe for healthy plants. And, if possible, position your pot so it is sheltered from strong winds and yet still gets good light.

Can I grow anything in a pot?

The short answer is 'yes', but remember that vigorous, large plants, such as oak trees and weeping willows will have a very limited life in a pot, because eventually the container they need will become so hefty as to be impractical. Wisteria is challenging too. If you can provide a large enough container and food, water and compost in sufficient quantity, you can grow almost anything in a pot. For large plants already growing in huge tubs, which are too big to pot on, scrape off about 10cm of compost each spring and replace with fresh to give them a boost.

What should I fill my pots with?

It depends what the plants are and how long they will be in the pot. For plants that are in a pot for a single growing season, for example summer bedding, a peat-free multi-purpose compost is ideal. Plants requiring a longer life in a container, shrubs and small trees for instance, need a compost that can hold on to nutrients longer than a multi-purpose compost, and one which is also heavier, to provide stability during windy weather.

For these, use an equal mix of peat-free multi-purpose and peat-free John Innes compost. The John Innes provides weight and longer-term nutrition, and the peat-free multi-purpose prevents it from becoming too compacted.

If the roots are poking out of the pot base, does the plant need a new pot?

Probably. At least, if you want to keep it happy. To be certain this is the case, knock it out of the pot and check that the compost is filled with roots. Sometimes roots push out of the bottom in search of water, if they have been kept too dry. If there is still plenty of compost that has not been colonised by roots, then the plant should be fine – just make sure you provide sufficient water and liquid feed in the growing season.

How do I stop pots freezing?

In the ground, roots are protected by the warmth of the earth, but in pots they're more exposed and can freeze – potentially damaging both plant and pot. In harsh winters, place pots against a wall, which acts like a storage heater to boost the temperature by a few degrees. Insulate the pot with bubblewrap, leaving the top uncovered to prevent rot due to poor air circulation.

Houseplants

INTRODUCTION

Houseplants can not only look modern and chic, but there are also health benefits to growing indoor plants. Greenery in our homes can make us feel calmer and increase our sense of well-being. Houseplants also help to purify the air by removing the toxins given off by furniture, cleaning products and electrical goods, which is especially useful during winter when we keep windows closed.

You might think that houseplants are hard to keep alive and healthy, but if you choose the right ones for the right spots, and give them the growing conditions they like, they're very easy to care for. To ensure you keep your houseplants happy and healthy, read our guide to some of the most common problems and follow a few simple rules.

'Choose the right houseplant for the right place (and that's critical) and you'll find your spirits will be lifted and your home will come alive. Just a few tender flowering and foliage plants will brighten up your life no end.'

Alan Titchmarsh

FIVE COMMON HOUSEPLANT PROBLEMS

Leaf scorch

Avoid standing your plants near a radiator or other heat source (their leaves will scorch and turn brown at the edges). If you see signs of leaf scorch, check that they are not in bright sunlight and if necessary find another spot where the plant gets bright light but not direct sun.

Leaf drop

Another common problem is forgetting to water, leading to leaves falling off. If this dry period is only brief, water well and the leaves will often grow back in a few weeks. This is common with ferns – any fronds that go completely crispy should be cut off at the base, but hopefully the plant will send up new ones.

Brown leaves

This can be due to overwatering or underwatering. If the leaves are brown but the soil in the pot feels wet, you may be overwatering. Overwatering, fluctuating temperatures and draughty locations are the main causes of leaf drop and browning. If you spot these symptoms, re-assess your watering regime and the positioning of the plant, and rectify if necessary.

Never water houseplants by the clock or the calendar. They'll dry out more quickly when the central heating is going full blast and when exposed to bright sun. Push your index finger into the surface of the compost: if it's as moist as a freshly wrung flannel, the plant will be happy. If it's drier than that, give the plant a good soaking, allow the pot to drain and then don't water it again until the compost feels dry. (Ideally use rainwater or cooled water from a boiled kettle.) Another way to judge when to water is just to pick up the pot, as dry compost feels light. Exceptions to the 'water when dry' rule are ferns and indoor azaleas, which should never be allowed to dry out. Cacti and succulents are also susceptible to overwatering, needing drier compost, but if kept too dry for too long they will begin to noticeably shrivel.

Houseplants keep dying

Dry air, caused by central heating, is one of the biggest killers of houseplants. When the heating is on, it's best to stand most houseplants on a tray of gravel, which should be kept damp to slightly raise the humidity of the surrounding air. Check regularly to make sure the gravel stays moist, but don't let plants stand in water. You can also mist the foliage occasionally with rainwater using a hand sprayer, but the effects of this are short-lived, and it's not good for hairy-leaved plants. Overwatering is the other common cause of death (see page 60).

Pest outbreaks

Most people wonder where pests come from – in many cases a plant already had them in small quantities and they went unnoticed at the time of purchase. Plants with papery foliage typically fall victim to red spider mite, while succulents and orchids are attacked by mealybug.

Pests such as red spider mites, scale insects and mealybugs all thrive in specific conditions, so read up and consider the location of your plant before trying to treat it. Once you're confident your plant is in the best possible spot, clean any pests off with your finger and thumb or damp cotton wool. If that doesn't work, you may need to resort to an organic soft soap spray.

ACTION PLAN
HOW TO KEEP YOUR HOUSEPLANTS HEALTHY

Feed your plants

If you don't feed your houseplants regularly, they tend to underachieve. You should start feeding regularly three months after you have potted with fresh compost and the nutrients have depleted. Use a liquid feed diluted in water as this is the best way to distribute it evenly. Follow the label instructions and if in any doubt dilute slightly more than the instructed amount to safeguard against damaging and scorching the plants. Most houseplants will not need feeding more than once every one to three months between March and September in their growing season.

Watering

You should aim to keep the compost just moist. Tap water is usually fine for most houseplants, but some, like carnivorous plants and azaleas, can be a little more demanding and benefit from rainwater. With the exception of some carnivorous plants, you don't want your houseplants sitting in water for days on end. To avoid this, raise the plastic inner pot off the bottom of the decorative outer pot with something like tile spacers – this will prevent plant roots sitting in excess water. Your plants will need more water during spring and summer when they're actively growing.

Keep up humidity for plants that need it

There are quite a lot of houseplants that benefit from high humidity – orchids, ferns and bromeliads are all good examples. Humidity can be easily increased with a hand-held water sprayer – it's good to give them a daily spray – and if you group plants together you will find they create their own humid micro-climate around them.

Repotting

Every few years, depending on the vigour of the plants, you should move them into slightly larger pots with fresh compost. The new pot should be only about 5cm larger in diameter – don't move them from an egg-cup to a bucket, as they'll sulk.

Dust leaves

Don't let houseplants get too dusty, as this reduces the light-absorbing capabilities of the leaves, making it harder for them to manufacture food and stay healthy. Wipe glossy-leaved plants with a damp cloth every few weeks, and dust hairy-leaved plants with a dry paintbrush.

Deadheading

To keep your houseplants looking good, snip off any faded leaves and flowers regularly. Turn the plants occasionally too, so they grow evenly rather than leaning towards the light.

'Foliage plants create a calm energy that radically improves the human condition, even in very cramped spaces. They also tend to be less demanding and many can be left untended for weeks at a time.'

Monty Don

PROJECT – How to repot a cactus

Cacti need repotting every few years, either because they're growing happily, or because they've become lop-sided and need straightening up. That said, their prickly nature can make this a tricky job, so you need to take care.

1. Fold up a piece of newspaper into a thick strip and wrap it around the barrel of the cactus, gripping the two ends together so they can be used to hold the plant steady.
2. Tap the cactus out of its pot and gently scrape away most of the old compost from around the rootball, trying not to disturb the roots too much.
3. Repot the plant in cactus compost, a gritty, loam-based compost that won't shrink when it dries out. Make sure the plant is central in the new pot and standing vertically.
4. Top-dress the surface of the compost with sharp grit to set off the plant and prevent it from being splashed with compost when you water it.

SPOTLIGHT ON
ORCHIDS

Many of us have been given a beautiful orchid as a gift – festooned with elegant flowers. But soon, the flowers fall off and you're left with bare stems and a few sad-looking leaves. And you wonder what you've done wrong. There's often a lot of mystery around growing orchids, but by following a few simple rules, you'll become an orchid whisperer in no time. Give your orchids the right care and you can enjoy healthy plants that flower for 12 weeks or more at a time.

~ **When watering** your orchids, imagine how they'd get water in the wild, as they cling to the branches of trees in a rainforest – the roots need a regular supply of moisture but should never sit in water. Either give a shot glass of water a week or soak in a sink of water for an hour once a week, allowing the pot to drain afterwards. They will need slightly less watering in winter.

~ **Try to find somewhere with plenty of natural light**
– orchids prefer bright but indirect light – but keep away
from south-facing windows and really sunny spots. If you
want to put your orchid on a windowsill, opt for east or
west facing. To make sure your orchids have the right
amount of light you may need to move them to a windowsill in winter and then a shadier spot for summer.

~ **Avoid putting orchids near draughts and radiators.**
They don't like temperature fluctuations. Place orchids on
a windowsill free from drafts or somewhere with a lot of
bright light. Orchids need a steady temperature – between
18–21°C is ideal for moth orchids but dendrobiums like it
a bit warmer, around 21–24°C.

~ **Use a specialist orchid feed**, which can be sprayed or
watered on. Apply this weekly each time you water, then
every fourth week revert to plain water to avoid a build-up
of salts. Avoid feeding orchids in the winter, but don't be
shy about feeding during the rest of the year and while
they are flowering.

~ **Deadhead regularly.** As the main flowers at the tip of
the stem fade, look very carefully for signs of nodes – little
joints that appear midway down the stem. Those nodes
want to flower too, so cut back the stem to just above the
highest node. This will encourage new growth. In next to
no time, you'll have a side branch with a gorgeous second-
ary display of blooms. Stubborn orchids can sometimes be

coaxed to re-flower by placing the plant in a place that's 5–10°C for four weeks and then returning it to a much warmer spot.

~ **Yellowing leaves** are perfectly normal and no cause for alarm. Don't cut them off, just pull gently and if they're ready to be removed they'll come off easily. Discard or put onto your compost heap.

~ **The leaves of your orchid need light**. You'll be amazed at how much dust can accumulate on plants in the home, which prevents the leaves from photosynthesising effectively. Remove dust with a soft, damp cloth, as part of your weekly care routine.

3 top tips

• **Do your research** before you buy, to ensure the plants will suit the conditions in your home. Then, with a little care and attention, you should be happy together for many years to come.

• **Move plants nearer** to windows in winter when there's less light available. It's also worth cleaning the glass inside and out, to maximise light levels.

• **If growing houseplants** on a windowsill, don't close the curtains over them at night, as this will trap cold air and chill your tender plants.

HOUSEPLANT CALENDAR

Spring

~ Feed houseplants between March and September, around every one to three months

~ Start watering more now that the growing season has begun again

~ Dust leaves

Summer

~ Feed regularly as your plant needs it

~ Cut off any browning leaves

~ Mist humidity-loving plants daily

~ Stand humidity-loving plants on a tray of moist gravel

Autumn

~ Stop feeding houseplants

~ Move plants away from radiators when the heating is switched back on

Winter

~ Keep houseplants away from draughts

~ Reduce watering

~ Don't feed houseplants over winter

Q&A
COMMON QUESTIONS
ABOUT HOUSEPLANTS

What houseplants should I avoid to keep kids and cats safe?

Most animals instinctively avoid poisonous plants, and children should be taught not to eat plant material from infanthood. Make your own risk assessment, especially where there are small children present. Don't just think about toxicity, though: allowing overhanging stems to tumble over the side of furniture within reach of small, tugging hands can also be very dangerous. Houseplants to avoid for cats include peace lily, dieffenbachia, Swiss cheese plant, devil's ivy, aloe vera, figs and euphorbias.

Can houseplants really filter the air?

It's widely accepted that plants can filter pollution from the atmosphere, reducing harmful chemicals such as benzene. Since NASA's tests on air-filtering houseplants in the 1980s, scientists have been comparing the influence of individual plants on air quality. Two common houseplants that score highly are mother-in-law's tongue (*Sansevieria*) and the peace lily (*Spathiphyllum*).

Hairy, rough or needle-like leaves are best for capturing pollutant particles, which are deposited on leaf surfaces or removed from the air as plants take in gases. While botanists still have more research to do in this field, it's clear that gardeners can make a vital contribution to air quality in our homes and cities, and to the health of their inhabitants.

What are succulents?

Plants that retain water in their stems or leaves, allowing them to survive in deserts and other arid areas. Though often treated separately, cacti are succulents too, storing water in their stems.

My houseplants are always dying. Which ones are easiest to look after?

Here are three suggestions for low-maintenance plants that are good for beginners. Cast iron plants (*Aspidistra*) are known for their tolerance of low light and neglect. Their relatively thick rhizomes and dark leaves enable these handsome plants to thrive in otherwise difficult locations.

Cape primroses (*Streptocarpus*) are good for people who can be on the forgetful side when it comes to watering. Their textured leaves and tender stems allow them to hold water for several weeks while still producing wonderful summer blooms in many shades of blue, purple and pink.

Zebra cactus (*Haworthiopsis fasciata*) is a sun lover well adapted to going for weeks without water. Originating from South Africa, it's the perfect companion for those seeking instant impact, even on the warmest of windowsills.

Weeds

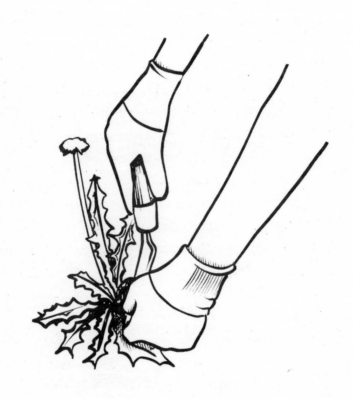

INTRODUCTION

There are no two ways about it – gardeners are obsessed with weeds. It's easy to see why: they compete with the plants we love, they appear in every available nook and cranny, and they interfere with the aesthetic appreciation of our well-cultivated plot. But what exactly is a weed? The precise definition is that a weed is a 'plant growing where it is not wanted', but to most of us a weed is the name we give to any British wildflower (mainly the ugly ones!) that invades our beds and borders. There are annual kinds that seed themselves freely, and there are perennials with tenacious roots that are the very devil to eradicate.

Of course, these days native wildflowers must be encouraged in all our gardens to benefit insect life, so cultivating a little tolerance is good for everyone. But some weeds can become a real nuisance – and should be your focus. The challenge for the organic gardener is to control weeds without resorting to chemicals, which, as far as the amateur is concerned, are soon likely to be phased out anyway. Instead of spraying with noxious fluids that can have a damaging effect on our environment and upset the balance of nature, it's better by far to look for alternative means of keeping on top of these plant invaders.

Different types of weed are a problem in different areas of the garden. On the veg plot it's usually annual weeds that are a nuisance – self-seeding and growing rapidly. In beds and borders, where the soil is disturbed less frequently, perennial weeds have the chance to get their roots firmly established. But even the toughest weeds can still be controlled by hand, without making a lot of extra work for yourself, if you go about it the right way. There's nothing you can't get rid of, over time, just by regular hoeing (and that applies even to ground elder and bindweed).

'How do you prevent weeding becoming an arduous chore? The big mistake is to keep putting it off. The longer you delay, the worse it gets. Conversely, early weeding is quick and easy – and rewarding. The garden always looks better when treasured plants are not being overrun, or rows of vegetables smothered.'

Alan Titchmarsh

FIVE COMMON PROBLEM WEEDS

Bindweed

A twining weed with thick roots and white trumpet flowers. It scrambles through border plants and shrubs. Pull out stems and fork out as much root as possible every time you see it.

Horsetail

It looks like little Christmas trees, but has very deep roots and can take over borders. It responds to regular pulling and even hoeing, but you have to keep at it for a long time.

Ground elder

Green elder-like leaves and white flowerheads. It forms a thick mat of roots and can force out other plants. Keep pulling out the thick white roots.

Stinging nettle

Tall plants with jagged-edged leaves that have stinging hairs. They make a good site for butterflies to lay eggs, but they are best confined to a single sunny corner of the garden. Fork out its roots and wear thick gloves.

Groundsel

A little green weed with clusters of small yellow flowerheads. It seeds itself about rapidly, so hand pull it or hoe out seedlings.

ACTION PLAN
FOR WEEDS

It is useful to distinguish between annual and perennial weeds. Annuals such as chickweed and speedwell complete their life cycle within a year and are a nuisance rather than a major threat. The ones to beware of are longer lived (or perennial) weeds such as couch grass, bindweed and ground elder. These are vigorous, persistent and regrow from even tiny pieces of root. Prevention is always better than cure, so try the following tactics:

~ **Mulch bare soil** with chipped bark to deter weed germination.

~ **Hoe frequently** when weeds are small, ideally on a dry, breezy day, to sever shoots from the roots. This kills annuals and slows the growth of perennials. Hand pull larger weeds and dig out perennial roots.

~ **Clear weedy ground** by covering with weed control fabric or thick cardboard to exclude light. Within a year or so, even the toughest weeds will be killed.

~ **Avoid using weedkiller** if possible, due to its negative impact on the environment. Alternatives include careful use of a weed burner to kill small weeds in paving and gravel.

If spreading perennial weeds invade clumps of perennial flowers, the only solution is to dig up the clump in autumn or spring and divide the plant, removing all weed roots before replanting. If creeping weeds spread from next door, sink a vertical barrier of polythene or lawn edging along the fence line to keep your side clear.

Weeds will grow in the gaps between slabs, in cracked concrete grouting, and in tiny traces of organic material trapped in crevices. Hand-weed using an old knife, or use a pressure-washer to blast out weeds and moss – spring is the best time to do this. Alternatively, use path weedkiller at the start of the growing season.

Grow cultivated plants close together in well-enriched soil, so they quickly form an impenetrable blanket over the ground, allowing little room for opportunist weeds. Bare earth is an open invitation to outsiders.

Try weed-suppressing membrane – a black, close-weave mesh that allows rain through but stops weeds growing. Cut slits in it to plant through and cover with bark mulch to hide it. This is best used among shrubs where regular access is unnecessary and for paths in woodland gardens.

Use groundcover planting, low-growing plants – often evergreen – that form a dense mat of growth and suppress weeds. These are useful for the front of beds and borders, and for shady spots under trees and shrubs.

'Weeding is not just about making your garden look good; it's also about improving the conditions for the plants you want to keep. If you do get overrun with weeds it won't only be tough for you, but also for your plants, as the weeds will soon start to steal water, nutrients and light, so it really is an important part of getting the best out of your garden.'

Adam Frost

PROJECT – How to hoe young weeds

A long-handled hoe makes short work of young weeds. But if you're weeding around and under plants, a long-handled hoe could easily cause damage so it's better to use a short one. A razor hoe with an angled blade is particularly good for close work. If you don't have time to give an area a full weed, snap off the weeds' flower heads to make sure they are never allowed to set seed, which at least helps in the long run. Hoeing beds on a regular basis, before weeds start to show, really helps to save time further down the line.

1. Hoe between ornamental plants in borders and crops in your veg patch, starting in the spring while weeds are still young.
2. Use the hoe to cut the tops of the weeds off. Pick a dry, breezy day so that the weeds will quickly wilt and die. Don't hoe too deeply as this may bring ungerminated seeds to the surface.
3. Leave the weeds to dry out on top of the soil, then pick them up and dispose of them once they are dry.
4. Repeat a few hours later to make sure you get any you may have missed. Mulch the ground after weeding.

SPOTLIGHT ON
PERENNIAL WEEDS

Perennial weeds fall into eight basic root types, each of which presents a different challenge to the gardener. Being aware of what you're dealing with, whether it's deep tap roots or short, spreading weeds, will help you take effective action.

~ **Tap roots** – this type includes cat's ear, dandelion, dock, thistle. The thick main root goes down deeply to survive winter and tap moisture from deep in the soil. Winkle out tap roots from lawns and paving with a daisy grubber or an old knife.

~ **Large rootstock** – includes black and white briony and pokeweed. These develop rapidly with a corky root, often more than 20cm under the surface. Dig deeply to remove the rootstock and all roots. Repeat every few weeks as soon as new growth appears.

~ **Dense mats** – includes grasses, nettles and yarrow. These weeds spread thickly under the soil surface to make dense, firmly rooted clumps. Best to dig them out with a strong spade or use a garden fork to lever out clumps. Small pieces take root easily if missed.

~ **Woody scrub** – includes brambles, sycamore and ivy spread by seeds, and quickly forms robust roots. Brambles and ivy also root at shoot tips. Dig out when young. For established woody shrubs, cut stems close to ground level and repeat for any regrowth to 'starve out' roots.

~ **Brittle roots** – includes bindweed and field bindweed, ground elder, creeping thistle and willowherb. These have wide-spreading root systems which break up readily and produce shoots from the smallest pieces. Sieve soil to remove roots. Keep cutting off the tops of these weeds to weaken roots.

~ **Deep-rooted weeds** – includes horsetail and Japanese knotweed. Roots can extend more than a metre down into the soil. Plants can spread widely close to the surface. It's best to clear the ground and cover the soil with old carpet or black polythene over the top to deprive these weeds of light for two years to weaken and kill the roots. If you find Japanese knotweed, there is legislation that covers its control and disposal, so seek professional advice.

~ **Bulbil type roots** – includes lesser celandine, oxalis, ransomes, Spanish bluebells, three-cornered leek. These pretty plants spread widely, but quickly outgrow their welcome. Propagated by small bulbils. Hoe off the leaves before flowering and repeat. Alternatively, cover soil with old carpet or black polythene from March to June to suppress growth.

~ **Creeping roots** – includes chickweed, clover, couch (twitch), creeping buttercup, daisy, selfheal, speedwell. Short, spreading weeds, with stems above or just below the surface, and most troublesome in lawns. Rake the lawn to lift creeping stems, then mow twice in opposite directions. Raise mower blades to leave grass longer and shade out weeds.

'I'm not sure there's any garden task that has a worse reputation than weeding, and thinking back to when I first started as a 16-year-old trainee, guess what the job was that I hated most? Yes, weeding! But today I'm more than happy spending a few hours weeding. This may sound sad, but there's something about this chore that I find really satisfying and therapeutic, and I love looking back at the ground I've worked on and seeing the difference it's made to my patch.'

Adam Frost

3 top tips

- **Hoe or pull** rather than dig, which would bring a fresh load of weed seeds up to the surface. Pulling is easiest when the soil is damp, whereas hoeing works best on a dry, breezy day.
- **Plant mixed borders** or naturalistic planting in a dense mosaic that leaves little room for weeds. Today there's much less desire for a 'perfect' green lawn, with many erstwhile lawn weeds, such as clover, now positively welcomed.
- **Remember the old** adage 'one year's seeding is seven years' weeding' and try not to let any weeds produce seedheads.

WEEDING CALENDAR

Spring

~ Hoe weeds from soil on dry days while they are small and easy to remove
~ Lay mulch to suppress any growing weeds
~ Fill planting gaps to stop weeds taking over
~ Tackle lawn weeds

Summer

~ Look out for perennial weeds – dig them out before they get established, making sure to get as much root as possible

~ Weed between crops as they will compete with your vegetables for nutrients

~ Take out weeds before they flower and scatter seeds

~ Look for any weeds hidden by tall perennials or shrubs

Autumn

~ Dig up any perennials that have been overtaken by weeds and remove before replanting

~ As plants die down, look for any remaining perennial weeds and remove all roots and stems

Winter

~ Do a last check for any weeds when you tidy up the garden for winter

Q&A
COMMON QUESTIONS ABOUT WEEDS

How often should I hoe in my borders?

Try to do it every week or ten days. In public gardens, staff tend to hoe the soil before they see any weeds; this might sound excessive, but it dislodges germinating seedlings so they dehydrate and die. It is also very quick and there's nothing to collect up and remove afterwards. It even works with perennial weeds: with their tops constantly cut off, they have no time to recharge their roots with carbohydrate, and so you literally starve the plants out.

Will mulch stop ground elder growing?

In a word, no. Perennial weeds like ground elder will have no trouble pushing up through mulch, and can even evade layers of weed-suppressing plastic material by running underneath and popping up round the edges. Your best defence is to fork it out regularly, or chop through new growth as soon as it appears. But don't give up on mulch. By excluding light from the surface of the soil, it prevents annual weed seeds from germinating, which saves a lot of weeding time.

How do I treat mind-your-own-business?

Mind-your-own-business is a shallow-rooted plant, so in borders dig it out and hoe regularly, raking up the bits so they don't re-establish. Don't put any remains in the compost, but instead bag them up and leave to 'cook' in the sun. If mind-your-own-business is in a lawn, don't rake as that will spread it.

I'm too busy to hand weed. What can I do instead?

Weeding can be made a lot quicker if you only fork out perennial weeds (they usually have deeper roots), and hoe annuals (weeds that live for a year or less). When done regularly, hoeing is quick and easy, as you'll catch the weeds when they're still small and, importantly, before they produce seeds. Keep the blade sharp and hoe on a hot day, so the weeds frazzle up on the soil surface. After weeding, you could mulch the soil with a 10cm layer of well-rotted compost or manure. This will reduce regrowth by deterring seed germination.

I'm new to this – how can I tell weeds from the plants I want?

Telling your weedlings from your seedlings is difficult at first, but you'll soon learn to identify your garden's main weed species. Try online weed identification charts. A couple of tips: when sowing seeds, mark where they are and ideally sow in rows. Anything outside that row is then clearly a weed. Try sowing a few spare seeds in a labelled pot each time you sow flowers or veg, to create a 'seedling reference library'.

Will I ever get rid of horsetail?

This is resistant to weedkillers, but you can control it by hand with regular routine gardening. If it's in your lawn, mow frequently, even in hot summers when the grass stops growing, as the horsetail will carry on and you need to keep chopping the stems. In borders and paths, hoe off the tops when they're buff-coloured fingers, before they get to the 'Christmas tree' stage. Within a couple of years, you'll just have the odd shoot or two.

Pests

INTRODUCTION

While many creatures are welcome in our gardens, those with voracious appetites for our prized plants can be disheartening. To control outdoor pests, an integrated approach is the most effective: encouraging natural predators, using biological controls if necessary and catching infestations early to avoid big problems later.

Above all, grow your plants well. Pay attention to the soil – enrich it with garden compost and well-rotted manure so that its organic content is high and moisture retention is good. Many garden plants fall prey to pests and diseases simply because they are hungry and thirsty. In the same way that we are more susceptible to disease when we are weak and undernourished, so are our garden plants. Grown well, in good earth, they are less likely to succumb to attack, and when a pest does strike they are more capable of surviving. Make sure you know what your plants look like when they are healthy. This may sound odd, but if you can tell when they look out of sorts it will ensure that you act speedily to counteract potential problems.

'Working with nature rather than against her is the way forward, and as someone who has not sprayed or dusted with chemicals for more than 30 years I can vouch for the fact that my garden is not overrun by pests and diseases, and that all manner of organisms – when allowed to do so – will achieve a natural balance that rarely gives rise to problems.'

Alan Titchmarsh

FIVE COMMON PEST PROBLEMS

Vine weevils

These are hard, black insects with elongated snouts. The white larvae live in the soil and attack plant roots, causing wilting and, in severe cases, death. Pot plants are most vulnerable. The adults damage leaves of shrubs and other plants. Strawberries, in both pots and soil, are often attacked. Go out with a torch at night to catch adults. Encourage toads and birds into your garden to eat the beetles and their larvae. You can also use nematodes to control larvae. Do not re-use compost from any pots that have been affected.

Aphids

Aphids are common sap-sucking pests, including green-fly and blackfly, typically found on new shoots, but they can occur anywhere. Look out for stunted or distorted growth and sticky honeydew with accompanying sooty mould. Aphids are visible to the naked eye. If you see white and fluffy infestations, these may be woolly aphids. You'll find them on most plants, but woolly aphids usually occur on the stems of woody plants, such as apple and pyracantha. There are a few ways of dealing with them. You can remove aphids by hand, rubbing them between your finger and thumb, or with a jet of water. Hang yellow sticky traps in the greenhouse. Encourage birds that will feed on aphids. Wasps, lacewings and hover-flies also help, so grow plants that attract these beneficial insects.

Slugs and snails

Needing no introduction, slugs and snails are the bane of UK gardeners. Look for irregular holes in leaves and slime trails. Some species bore holes in potatoes. They will attack most plants. Lure them into beer traps or upturned halves of melon and grapefruit, or use barriers like copper rings, ash or eggshells. Copper collars placed around the plants as the shoots emerge from the ground are effective. Enrich the earth with organic matter to encourage faster, more vigorous growth. A good, simple method, although time consuming, is just to hand pick them off the plants after dark. Regularly cultivate soil and disturb leaf piles to expose eggs. You can also control slugs with a nematode, but this needs to be applied at the correct time, so check the instructions. It's a good idea to try to attract predators such as birds, hedgehogs and toads.

Caterpillars

Cabbage caterpillars are laid by large and small white butterflies. They eat the foliage and leave behind nasty black frass (poo!). Look out for ragged holes in leaves. Cover the plants with crop protection netting to prevent the butterflies laying eggs. It is a wider mesh than the netting used to repel carrot fly. The caterpillars of large white butterflies are yellow, black and hairy, while the caterpillars of small white butterflies are pale green with short hairs. Also look out for cabbage moth caterpillars, which are not hairy and a yellowish or brownish green. You can also try removing the caterpillars by hand or using a nematode. Encourage garden birds as well, as they will eat the caterpillars.

Whitefly

These are small, active sap-sucking insects that breed rapidly. Whitefly colonise the undersides of plant leaves, especially tomatoes, weakening them and secreting honeydew. Look for leaf yellowing and distortion. Whitefly is easily visible. You'll find these on many indoor plants such as cineraria, fuchsia and poinsettia as well as greenhouse veg such as aubergines, cucumbers, tomatoes and peppers. Plant French marigolds (tagetes) in the same soil as the tomatoes. The pungent smell given off by the marigolds seems to be effective in discouraging attack. In greenhouses, try hanging sticky yellow plastic sheets to trap the pests.

ACTION PLAN
HOW TO PREVENT
PEST ATTACKS

~ **Avoid problem times for pests.** Pests are attracted to weak plants, because one of their roles is to recycle debris and decay. You can help veg grow into strong and sturdy plants by sowing seeds at the best time of year for that crop. For example, sow broad beans in February, so they have plenty of time to grow before their natural flowering time in early summer. In contrast, runner beans are tender, so wait until May to sow seeds, to avoid the risk of frost damage or checks to their growth due to cold weather. And by sowing rocket in August, rather than in spring, you miss the flea beetle season.

~ **Cover susceptible plants** with insect-proof mesh. Try keeping aphids in check by spraying leaves with water, until predators arrive naturally.

~ **Use a natural spray.** For persistent aphid problems try this potion: blend three chillies, three garlic cloves and a cup of mint leaves with six cups of water and a little eco washing-up liquid, then spray onto affected plants. One application should be enough, or you risk killing predators, such as hoverflies. Take care and wash your hands thoroughly after use.

~ **Introduce a biological control,** an organism that will feed on or actively discourage the multiplication of another organism, such as a predatory nematode. Such predators are available for many pests, from whitefly and red spider mite to slugs and vine weevils. These controls have their limitations – they are usually only active and therefore effective above a certain temperature, which governs the time of year at which they can be used.

~ **Look for plants that are resistant to attack** – in the case of slugs, try hairy plants or those with tough, leathery foliage such as stachys or pulmonarias. Some plants, such as foxgloves and euphorbias, have toxic leaves.

PROJECT – How to prevent earwig damage

If the petals of your prized blooms are being eaten, earwigs are likely to be the culprits. Dahlias and chrysanthemums are particular favourites of these pincer-wielding insects, which like to devour the soft tissue of petals, rather than tougher leaves or fruit. Earwigs feed from late spring right through the summer months. After hiding in crevices or among debris during the heat of the day, they climb up flower stems at night to feast on petals. Their mixed diet also includes small insects and their eggs. If you spot earwigs on a fruit tree, they're feeding on aphids.

It makes sense to trap problem earwigs in a pot and move them from your flowers into the fruit garden, where they can help to control aphids rather than nibble on your best blooms. Check the traps every morning and shake any earwigs out of the straw. Release them elsewhere in your garden, then repack the pot ready for the next night's catch of petal munchers. This is how to make an earwig trap:

1. Insert a cane into the ground among your affected blooms so the top is at flower level. Wrap string round the cane just below the tip and knot in place – this will form a barrier and stop the pot sliding down.
2. Using a knife, carefully pierce a small cross in the base of a 9cm plastic pot. Then stuff the pot with plenty of straw, tightly enough to ensure that when the pot is held upside down the straw doesn't fall out.
3. Make a hole through the centre of the straw with your fingers, so that when it is placed upside down onto the cane, the straw doesn't become further compacted into the base of the pot.
4. Gently press the pot down over the cane. The cross in the base should flex enough without splitting to allow the pot to fit firmly onto the cane and sit securely above the string buffer.

SPOTLIGHT ON
LARGE PESTS

It is not only the infestations of insects and fungi that trouble the gardener, but also larger creatures on two legs or four. These are no respecters of organic gardening, so natural 'biological' control is not an option. These blighters really give us grief, and lovers of nature find that the prospect of dealing with them brings on a crisis of conscience. But if we can at least minimise their visits, we can continue to enjoy guilt-free gardening.

~ **Mice can wreak havoc** with seedlings, newly sown seeds – especially peas and beans – and bulbs. Position chicken wire over bulbs at planting time to prevent their predations. Holly leaves that prick their noses can be laid in seed drills, but they will also prick you, which is irritating. Set humane traps (mice love peanut butter) and release the trapped mice where they can do little harm.

~ **Brassicas are Michelin starred** as far as pigeons are concerned, and wide-mesh pigeon-proof netting is the only effective answer. Fix it securely over a bamboo framework and raise the homemade cage as the crops grow.

Walk-in fruit cages will make sure that you, rather than the birds, harvest raspberries, currants and the like. Birds will also take flower buds from blossom trees, but here there is little you can do to avoid their plundering.

~ **To deal with foxes,** the most important thing to do is make sure that all household waste is securely shut away. Don't leave bin bags of rubbish lying around or you will come down in the morning to discover your garden looks like the local tip. Ultrasonic devices are available, as are chemical deterrents, but these are of varying effectiveness, and sprays need to be regularly re-applied. Many of them are based on strongly scented chemicals that you might find almost as offensive as the fox. Make sure that places of shelter (under decking and verandas) are sealed off with wire netting to avoid easy access. The key is to make sure that your garden offers as little attraction to foxes as possible – if there is nowhere to shelter and no food, they are likely to move on to places where they can find both.

~ **Squirrels** – to deter their digging activities lay wire netting just under the surface of the soil after planting bulbs in beds or containers. Squirrel-proof bird feeders (with an outer mesh cylinder that squirrels cannot penetrate or that tilts and closes when larger pests land on it) will ensure birds not squirrels get the food.

~ **Deer** browse trees and shrubs, eating the foliage and young stems within their reach. They also rub their developing antlers on young tree trunks (a process known as fraying), which can ring-bark the trees and cause death. The only guaranteed way to prevent deer damage is either to fence the entire property with 3m-high fencing, or to fence each susceptible young tree with a 2m-high cylinder of stock fencing.

~ **Rabbits** will eat plants they find tasty, but they will also ring-bark young trees. To keep them out of your garden you need a sturdy wire netting fence 1m high above ground and 45cm below ground to prevent their burrowing underneath it. Spiral plastic rabbit guards placed around newly planted trees are an effective and more economical solution. An online search will reveal long lists of plants that are less attractive to rabbits and deer.

~ **Moles** – the bane of any gardener with a decent lawn. Use the molehills as an ingredient of homemade potting compost. That's the positive bit. If you live next to farmland, they will always come back, but if your soil is on the heavy side try a 'mole mover', which is a battery-powered spike pushed into the ground and which sends out repellant vibrations that make the mole back off from whence they came. These devices are less effective on light, sandy soils. Powdered mothballs worked into the molehills will not be to their liking (whole mothballs are simply pushed out again). Children's windmills stuck into the hole are said to repel them, but then your garden will look like Blackpool on a bank holiday.

~ **Cats** – it's tempting just to say get a dog and leave the advice at that. Why is it always next door's cat that uses your garden as a loo? *Plectranthus ornatus* (commonly known as 'scaredy cat' or *Coleus canina*) is said to repel them, as are rue and lavender, but you can't plant them everywhere. Various chemical preparations are available, but they seldom last for long. You can also buy ultrasonic devices that are said to ward them off. Twiggy peasticks laid across newly sown seedbeds will deter them from scratching up the earth to perform their ablutions, as will short lengths of garden cane pushed into the ground vertically every 20cm or so during germination.

~ **Badgers** – of all the mammals, this is the one that can wreak the most havoc, digging huge holes in lawns and burrowing under hedges for access. They are nocturnal and are almost always looking for food. Even the most robust wire fence buried in the ground will often prove no obstacle. If badgers persist in entering your garden looking for a meal, you could try giving them one: placing a tin tray containing things like peanut butter, whole peanuts, dry dog food, fruit and honey close to their point of entry in the hope that they will not feel the need to look further into your garden.

3 top tips

- **Buy healthy plants** from reputable nurseries and garden centres. Be wary of cheap foreign imports, which are the source of many pests and diseases.
- **Take preventative action** against slugs and snails. An important step is to reduce potential slug and snail habitats close to vulnerable plants – young seedlings, salad leaves and brassicas are particularly at risk. Dark, damp places are havens for slugs and snails, so assess your plot and consider how you can minimise these.
- **Encourage a wide** range of birds into the garden by putting up feeders and nest boxes as well as filling a bird bath. Grow berrying plants and leave seedheads in the autumn. They will soon start feeding on pests as well, such as aphids and larvae.

'I find that a surface mulch of compost on undug soil gives plants a head start against slugs in the spring. As the season progresses, keep the growing area scrupulously tidy by removing all weeds and any diseased and yellowing leaves, so that mollusc numbers cannot increase as they have nowhere to hide by day.'
Charles Dowding

PEST WATCH CALENDAR

Spring

~ From spring you can apply nematodes if you're going to use them – check individual types for timings

~ Check for aphids and control infestations before they get out of control

~ Check for pests in the greenhouse – rub off any greenfly from plants' new buds

~ Look out for slugs feeding on new shoots

~ Place twigs over bare soil to prevent cats using your borders as a litter tray

~ Keep your veg crops well watered to prevent flea beetles making holes in the leaves of cabbage, broccoli, rocket and radish seedlings

~ Remove blackfly from the tips of broad beans

~ Look out for telltale holes in viburnum leaves, indicating viburnum beetle – pick off infested leaves

~ Pick vine weevils off plants – each one can lay thousands of eggs in spring and the resulting grubs cause damage to plant roots.

Summer

~ Squash or wash off aphids on shrubs such as roses

~ Check regularly for slugs and use barriers to prevent them feeding on plants

~ Look out for lily beetles and their eggs. Remove the beetles by hand to prevent damage to leaves

~ Check greenhouse plants regularly for pests including aphids, red spider mites and whitefly which can multiply rapidly in warm weather

Autumn

~ Leave seedheads on plants to help attract birds to your garden to eat the pests
~ Prune out the shoots of bay trees that were damaged by bay sucker pest during the summer. The leaves will look yellow and curl inwards
~ Put chicken wire over containers of bulbs to stop squirrels stealing them
~ Make an insect house to encourage beneficial insects such as ladybirds and lacewings

Winter

~ Rake up fallen leaves in borders that could be harbouring slugs and other pests
~ Hang bird feeders near borders to encourage hungry birds that may also feed on overwintering pests
~ Go on a snail hunt – in winter snails hibernate under the cover of foliage, stones, paving slabs and behind climbers. Pick them off and take them at least 20m away if you don't want them to return in the spring

Q&A
COMMON QUESTIONS ABOUT PESTS

How can I stop squirrels digging up my bulbs?

Grey squirrels are observant and persistent. Like birds they're initially scared by fake predators, such as cats, birds of prey and snakes. But not for long, even if these are moved thrice daily. A yappy dog may help, but as soon as it's back indoors, the squirrels will return. Proprietary 'discouragements' – sprays and dusts based on garlic, pepper, soap or chilli extracts, even old after-shave – can work temporarily, but do not last long as weather degrades them.

A movement-operated floodlight, buzzer or sprinkler, or better all three, might just work. However, the best answer is to use small-gauge galvanised chicken netting pegged over the bulbs and buried just sufficiently to be out of sight. The leaves and flowers can grow up through it, but the squirrels cannot dig up and eat your bulbs.

What can I do to get rid of rosemary beetles?

They can be a real menace! It is worth remembering that as well as rosemary and lavender, these beetles and their larvae will also damage sage and other closely related plants, so check those, too. On the whole, the beetle rarely does enough damage to have a significant impact on the plants' health.

Try to collect as many beetles as possible by hand – you can speed things up by spreading an old sheet beneath the infested plants and then shaking them. This way, you can collect significant quantities of these pests. There are a number of natural predators of the rosemary beetle including ground beetles (which eat the larvae), frogs and toads, so ensuring that your garden is a good place for these creatures to live will also mean that you have plenty of help.

Pests are eating my veg seedlings. What should I do?

Horticultural fleece creates a useful barrier against airborne pests, such as whitefly, carrot fly, aphids and pigeons, but it can't stop ground-dwelling pests or those that are already established in the bed.

If it is not slugs, the caterpillars of several different moths, called cutworms, could be the problem. You can find caterpillars at any time of year, so regularly poke around the affected plants to find them and remove by hand. Birds, ladybirds, wasps and ground beetles are their natural predators. Keep your crops well watered and weed-free, use biological-control nematodes and cover with fleece in summer.

How do I keep vine weevil off my dahlias?

Vine weevils are all female and need no males, so just one can start a new infestation unaided. They can crawl to almost anywhere, but they cannot fly and it's doubtful they can swim. Thus, moating each of your containers is the most effective measure to stop them spreading and multiplying.

Find sufficient 'saucers': proper pot plant ones, soup plates or plastic lids that can fit under your containers. Fill these with water and stand the containers *above* them, each resting on three pebbles or similar.

Each container is now safe from re-infestation or exporting adults but will need individual inspection to remove any adults and grubs from continuing to feed and breed in it. Having removed any found, you could then apply the commercially available nematode to kill any missed or hatching from eggs. This is best added in warm conditions.

How can I stop badgers digging up my lawn at night?

Badgers, which dig up lawns looking for worms and grubs, can be difficult to deter as they go round barriers or dig under them. Try to block all access to your garden – it is easy to tell their entry and exit points – and don't feed wildlife or leave food for pets outdoors. The RSPCA website has a wealth of information on sharing your garden with badgers. To discourage them from digging up your lawn, you could try leaving food out for them, such as carrots, potatoes and sweetcorn, but don't give them too much or they will become dependent. Often they are looking for chafer grubs to eat. These can be controlled with the biological control nematode *Heterorhabditis bacteriophora*, which is watered into the lawn. Controlling badgers is not easy, and it is illegal to do anything that harms them or their sett.

What is box tree caterpillar?

Box tree caterpillars are the larvae of a moth, *Cydalima perspectalis*, which lays its eggs on the undersides of box leaves. The moth is native to east Asia and is thought to have arrived in Europe in 2007. They are now a major problem, especially in the south east of England. Symptoms include cobweb-like webbing, droppings and die back, which look similar to box blight.

After around a month, the caterpillar forms a chrysalis which emerges as a box tree moth, which then mates, perpetuating the cycle. Box tree caterpillars can be a problem from spring to autumn, producing multiple generations. The caterpillars overwinter among box foliage, resuming feeding the following spring.

Young caterpillars are a greenish yellow with black heads; the older caterpillars are the same colour but also have black and white stripes. Try removing them by hand if numbers are small. Check deep within the plant. You can also prune out the stems covered in webbing, using secateurs. If the caterpillars are taking hold, a biological control that contains the microorganism *Bacillus thuringiensis* is said to be effective. Treatment needs to be repeated several times across the season, when the temperature is at least 15°C. Spray thoroughly, coating both sides of the leaves so it penetrates deep into the plants.

Diseases

INTRODUCTION

Plant diseases are malfunctions caused by organisms such as fungi, bacteria or viruses. Many diseases thrive in the warm months, and some are made worse by dry weather. Look out for symptoms on leaves, stems, flowers and fruit. Finding signs of disease can seem daunting, but many problems can be dealt with if caught early. Take action when you spot symptoms, and you'll give plants the best chance of recovery. As with pest problems, giving consistent care to plants to keep them healthy will improve their chances of avoiding disease. Some of the most common diseases include mildew, black spot and rust, but it's also important to look out for more serious problems such as honey fungus and leaf curl. Bear in mind that some diseases are harder to treat than others and you might not always be able to solve the problem. In these cases, it's best to get rid of the plant, take measures to prevent spread of the disease and invest in a resistant variety (if available) next time.

FIVE COMMON PLANT DISEASES

Powdery mildew

This is a fungal disease. Unlike many fungi, powdery mildew strikes in dry weather, stressing already weak plants. Look out for powdery material on leaf surfaces and flowers. It can cause leaf distortion, and affected leaves can die. Many plants are affected by this, such as acanthus, cherry laurels, hydrangeas, lonicera, phlox and roses, and vegetables including courgettes and peas. To control it remove all affected leaves and control weeds that harbour the fungus. Avoid drought stress by mulching plants and watering regularly. When watering, avoid getting the leaves wet.

Box blight

This fungal disease causes box leaves to turn brown and fall, and can lead to black streaks on young stems. Dead patches should be cut out and burned, along with all the fallen leaves at the base. If the entire plant is affected, it's often best to remove and burn the whole lot. Avoid clipping box when it is wet and the disease is more readily spread. Some new varieties, such as 'Heritage', 'Skylight' and 'Renaissance', are resistant to attack. Also consider plant alternatives such as *Ilex crenata* and *Euonymus* 'Jean Hugues'. Symptoms of box blight are similar to those of box tree caterpillar (see page 106).

Rust

This is another fungal disease – an infection that damages leaves, weakening plants.

Look out for orange or brown spots on the lower surfaces of leaves, yellow patches that appear on the upper sides of leaves and dying leaves falling off the plants. Many ornamental plants are susceptible, notably hollyhocks and roses, as well as fruit trees, raspberries and leeks. Remove and destroy any infected leaves, and try to grow rust-resistant varieties if possible.

Clematis wilt

A common fungal disease of clematis that, in some circumstances, can prove fatal. You'll see the upper leaves and stems wilting and turning black. Wilt quickly spreads down the plant. Plants may survive and re-grow. Remove and destroy affected material immediately. Large-flowered clematis hybrids are prone to it, while small-flowered hybrids and species such as *C. alpina* and *C. montana* are usually resistant. If this is a problem in your garden, grow resistant varieties. Clematis need a cool, damp root run so plant new specimens deeply and cover the base with an old pot or slate.

Viruses

These cause the twisting and disfigurement of leaves and shoots, as well as unwanted variegation – yellowing and streaking. Alas, there is no solution to plant viruses. Dig up and burn affected plants. Encourage birds and predatory insects to control the aphids that spread viruses.

'Thanks to a combination of climate change, globalisation and the free movement of goods, new plant diseases are arriving on our shores at an unprecedented rate. Some kill plants outright, while others simply disfigure them. Still, with good garden management and careful plant selection, it's certainly possible to lessen the impact of these diseases.'

Nick Bailey

ACTION PLAN
HOW TO HELP PREVENT DISEASES

~ **Keep tools clean.** Diseases can be spread from plant to plant by dirty blades of pruning tools.

~ **Choose resistant species** of plants such as roses, box and phlox.

~ **Increase soil aeration.** Healthy garden soil should comprise around 25 per cent air. Add lots of organic matter, which opens up the soil's structure.

~ **Use mycorrhizal fungi.** This may be effective in overcoming rose-replant disease because the fungal strands that attach to the plant roots effectively increase the plant's root zone, meaning it can extend beyond the affected soil.

~ **Many soil-borne** diseases are fungal in nature, which means they are spread by spores, so avoid relocating soil to different areas. If you have plants affected by clubroot, take care not to transport affected soil around your garden on your boots and tools.

~ **Clean your greenhouse** inside and out during winter. Good hygiene is an essential weapon in the war against pests and diseases.

PROJECT – How to deal with rose black spot

This is one of the most common rose diseases. Black spot is just that – large black spots that eventually coalesce and cause leaf drop. Plant roses with thick, leathery leaves that are naturally resistant to diseases such as black spot and mildew. Rugosa and old gallica roses are particularly healthy – try the rugosa rose 'Blanche Double du Coubert' or gallica 'Charles de Mills'. Grow plants in rich, moisture-retentive soil (this disease is most prevalent in poor, dry earth).

1. In autumn, collect and burn leaves infected with black spot.
2. In winter, prune out any stems that are affected.
3. In late winter, spread a thick layer of mulch around the base of any rose affected by black spot. This will help prevent rain splashing soil-borne spores on to new growth in the spring.
4. Finally, when buying any new roses, look for a variety that is resistant to black spot.

SPOTLIGHT ON
SOIL-BORNE DISEASES

~ **Soil is the lifeblood** of any garden. It provides nutrients, water and root anchorage to plants, but lurking within it are a host of sub-surface scoundrels. These are essentially soil-dwelling fungal and bacterial organisms that can have devastating or fatal consequences for your plants. Probably the four most common forms are rose-replant disease, clubroot, honey fungus and phytophthora root rot.

~ **Rose-replant disease** affects roses, apples and other plants in the Rosaceae family, when they are planted into a spot where the same type of plant was previously growing. Symptoms usually occur in the first year and include poor growth, yellowing leaves, no flowers and occasionally even the death of the new plant.

   The problem is thought to be caused by a build-up of fungal and bacterial pathogens in the soil. The original plant would likely have coped with them, due to its size and maturity, but a young, newly planted specimen has no such resilience. So avoid replanting in the same spot. If you have no alternative, add mycorrhizal fungi to the planting hole and replace as much of the soil with fresh as you can, ideally 50 litres or more.

~ **Clubroot** is a fungus-like relative of slime mould that affects brassicas such as cabbages, turnips and radishes. Their foliage becomes stunted, gnarled and often turns purple, but the main damage is below ground, where roots become swollen and contorted. There is no chemical control, although adding lime to the soil will knock it back a little.

Prevention and avoidance are your best bets. This means buying only guaranteed clubroot-free plants, and if clubroot is already present in your garden, avoid growing brassicas in the affected area. Instead, keep young brassicas in pots for as long as possible or grow them in raised beds filled with unaffected soil, and choose clubroot-resistant varieties.

~ **Honey fungus** is another widespread killer, attacking a range of woody plants and herbaceous perennials. Symptoms include pale leaves, dieback, early autumn colour, bleeding from the trunk and clusters of honey-coloured toadstools. There is no chemical control, but you can halt its spread by inserting a rubberised barrier vertically into the soil to about 50cm deep. Remove and destroy affected plants and their roots. This rampant fungus can attack most woody plants, but pittosporum, chaenomeles, box and tamarisk are less susceptible.

~ **Phytophthora root rot** is a fungus relative that attacks everything from herbaceous perennials to bulbs and woody plants. Poor growth, yellowing and dieback are the first symptoms, at which point you need to look below ground for roots that are rotting, soft and brown or black inside.

Smaller roots may have died off. Remove and destroy affected plants and improve soil drainage. Opt for plants that show more resistance, such as abelia, echinacea, ginkgo and liquidambar.

3 top tips

- **Stop diseases spreading** by putting infected material in garden waste bags – don't compost it. Although diseases can't be cured, some can be treated with fungicide.
- **Select disease-resistant varieties** wherever possible and buy healthy plants from a reputable source.
- **Adopt good gardening techniques,** such as watering onto soil rather than foliage to minimise disease spread, and spacing plants out to improve air circulation.

DISEASES CALENDAR

When it comes to plant diseases, focus on seasonal tasks that help keep plants strong and healthy and good gardening techniques that help prevent the spread of disease.

Spring
~ Inspect any new plants to make sure they are healthy before planting
~ Feed plants to keep them healthy and less prone to disease. Diseases can attack plants that have been under stress. Lack of water and food can cause stress in plants

~ Keep pricking out seedlings as soon as they get their first true leaves, to avoid overcrowding and fungal diseases

~ Water new plants regularly until established

Summer

~ Look out for diseases during warm weather, as this is a key time for problems such as mildew, rust and black spot

~ Water plants that need irrigation regularly, especially in hot weather – this includes new plants and those in containers and hanging baskets

~ Collect or remove leaves that are showing signs of disease and burn to prevent spread

~ Don't compost diseased material – burn it if possible or bin it if not

~ Keep plants well watered when growing under glass

~ Prune plum trees in dry weather when silver leaf fungus is less prevalent

~ Sweep the greenhouse regularly to reduce debris that can harbour diseases

Autumn

~ Gather fallen leaves from diseased plants and destroy them

~ Inspect tender plants for signs of disease before bringing them undercover

~ Mulch around plants in borders

~ Clear crops from the veg plot but don't compost any diseased material

Winter

~ In late winter, bin hellebore foliage marked with black blotches to limit the spread of leaf spot disease

~ Remove any foliage on pansies affected by downy mildew

~ Remove any diseased branches when pruning to stop the spread of disease

~ Check for rot on stored bulbs and tubers

~ Remove any yellowing leaves from overwintering plants in the greenhouse to prevent fungal diseases

Q&A
COMMON QUESTIONS ABOUT PLANT DISEASES

What is botrytis?

Botrytis is a catch-all term encompassing many related fungal diseases that tend to strike in areas of high humidity and low air movement.

My hydrangea's leaves are discoloured – what should I do?

Cercospora fungus is the most common cause of leaf spotting on hydrangeas. Others, including powdery mildew, may also lead to purple-brown foliage. Hydrangeas enjoy moist soil, so regular watering and a good bulky organic mulch around the entire root area will help to improve the growing conditions, making your hydrangeas less susceptible and better able to cope with any infection. Moist roots and dry air are also key to reducing powdery mildew. Fungal diseases are more likely if plants are too close together, so when planting new plants, give hydrangeas a bit of space to improve air circulation around them.

Why have my daffodil bulbs rotted?

They could be affected by basal rot, which is one of the most common diseases in daffodil bulbs. It's caused by a fungus in the soil. If you notice any fungus on stored bulbs, or the foliage of your daffodils turns yellow early in the season, this might be the reason. Don't plant any rotting bulbs.

What is peach leaf curl?

If you notice curling, puckered leaves in early spring on fruit trees such as peaches, ornamental cherries and apricots it could be peach leaf curl. This is caused by a fungus that infests new leaves in the spring. The fungus feeds on young leaves and distorts them. This can lead to flowers and fruit falling off the tree as the leaves are unable to make enough food.

Trees can recover and produce a second flush of growth – the best thing to do is collect any infected leaves from the ground and burn them to prevent further spread. If peaches are trained against a wall, cover with a polythene sheet in late winter (leaving the ends open for pollinating insects) to prevent the fungus developing. Keep plants healthy by mulching around the base of trees with garden compost and watering in hot summers. Avoid overfeeding with nitrogen fertiliser.

Garden
Design
Problems

INTRODUCTION

Every garden has its own unique challenges and charms, but there are key principles that can be applied to any plot to pep it up. These are simple and can make a huge difference to an established garden or a space being created from scratch.

Whether you have a blank canvas or already have borders in place, making changes can be challenging – but they can also be very rewarding. So if your garden doesn't suit your needs, has become tired or too familiar and no longer fills you with joy, then set about making it lift your spirits. Even minor changes will make a huge difference. Do bear in mind that you're bound to make some mistakes along the way – we all do – but remember that nothing is set in stone (except an expensive patio!). Plants that have been put in the wrong place will soon become apparent and can easily be moved.

Look at the views from your windows – the kitchen especially – and assess if your current garden looks pleasing from the most frequented spots. Whether your garden is old or new, analyse what you don't like about it and set about making amends.

FIVE COMMON GARDEN DESIGN PROBLEMS

Dull boundaries

Bald fences are easily remedied if there's soil at the base. They can be equipped with horizontal wires at 45cm intervals and climbers such as clematis and roses trained up them. If your fence is alongside a patch of concrete and there's no earth, try using containers. Size is key here. Climbing plants need plenty of sustenance at the roots to ensure they are fuelled for their journey up and along a fence or screen. The pot, tub or trough into which the climber is planted needs to be a generous size (minimum 40cm x 40cm) and filled with good compost, such as peat-free John Innes, to offer stability as well as nourishment. Don't grow wisteria as it needs a deep root run. Instead, plant bright-flowered clematis, star jasmine (*Trachelospermum jasminoides*) or firethorn (*Pyracantha*). If the fence receives plenty of sun, set a row of troughs at the base and plant climbing nasturtiums and runner beans and enjoy the bright flowers and the tasty crops.

Lack of year-round interest

Anyone can have an interesting garden in summer, but what about the rest of the year? A skeleton framework of evergreens coupled with a simple but pleasing overall design of beds and borders works well in a small space. Topiary is perfect – obelisks and orbs of yew and box throw moving shadows as well as creating statuesque shapes. Don't overdo evergreen shrubs, or your garden will begin to feel like a Victorian graveyard, but a scattering of them – like choisya, which has white blossom and shiny leaves – will ensure your garden does not simply disappear when autumn comes and the leaves fall.

And you know how you always shoot off to the local nursery or garden centre in May and June and come back laden with instant colour? Try visiting once a month throughout the year and bringing something home that is in flower or at the height of its interest. That way, you will continue to have at least something interesting in your garden all year round.

Lack of privacy

Small gardens tend to be overlooked by surrounding houses. Anyone who has read the horror stories of neighbourly disputes knows the perils of Leyland cypress hedges. The advantage of this thick evergreen hedge is that it is fast growing. The disadvantage is that it does not stop until it reaches 25m or more, plunging the land around into eternal gloom and turning the earth below dry and root-ridden, making it difficult to grow other plants. But clip it regularly – at least once a year in late summer – and you'll restrict it to a dense hedge that is under control.

If you are prepared to wield your shears or hedge trimmer annually, don't disparage hedges. They are wildlife friendly compared with fences, which are bald, boring and subject to being blown over by winds. Hedges are more effective at filtering the blast. Grow one to around 2m high and you'll be safe from the gaze of neighbours. If a fence is already in place, add a 45cm-high strip of trellis along the top so fast-growing climbers can be threaded through for extra privacy. A well-placed small or slow-growing tree (birch or crab apple) will prevent neighbours from looking down on your sitting area. A canvas shade sail supported by stout posts achieves the same effect on a temporary basis, and a criss-cross pattern of hedges or woven screens can also create secluded areas.

Ugly features

It is all too easy when designing a small garden to take little account of the less aesthetic aspects of domesticity – the dustbins, the shed and the washing line. But they are essential to daily life. Washing lines are never attractive – the secret is to ensure that they can be easily taken down and stored when not in use. Whirligig versions look less offensive if they are positioned in the centre of a circular lawn or – better for all-weather access – in a circular patch of paving. If set in the lawn, make sure the hole is kept clear of grass, or you will be forever searching for it. Better still, remove them when not in use and store in the shed, or install a retractable line – they disappear to nothing.

It's a good idea to erect low hazel or wattle screens alongside dustbins to screen them from view. And painting garden sheds in attractive or sympathetic colours really does improve their appearance, rather than settling for the lurid orange stain which all of them seem to sport when they arrive.

No storage

It can be tricky to include storage space in a small garden. Transform your shed by making use of the walls and ceiling by attaching hooks and nails to hang up tools, freeing up floor space. Add shelves to hold plant pots and fertiliser, and use baskets or crates to keep netting and fleece tidied away. Look out for old filing cabinets and vintage cupboards to keep all your odds and ends organised. Another option is to combine seating with storage – install benches with seats that can be lifted up to store things like children's garden toys and barbecue tools.

ACTION PLAN
HOW TO PLAN YOUR GARDEN

~ **Make seating areas as large as possible.** Once your table and chairs are on a patio or decking area, the space will feel a lot smaller and you will need space to move chairs in and out.

~ **Use focal points to create impact.** Objects or small trees will draw you on and make you want to explore. This could be a large container filled with seasonal planting or a small fountain or water feature, or even a striking plant in a border. This will draw the eye to certain points of the garden, such as the end of a path or corner.

~ **Make your shed a feature.** With a bit of planning and inspiration, a shed can offer a whole new space for living, playing and even growing. A simple coat of paint can lift a shabby shed instantly. Try contrasting colours, adding stripes or bright tones to draw the eye and make it stand out.

~ **Add a green roof to your shed or bin shelter.** A roof, no matter how small, is a great new planting opportunity, on what would otherwise be a barren space. Covering a roof with plants will soften the appearance of a shed or shelter, helping it blend with the greenery around it. A green roof will also provide insulation and offer shelter for wildlife too. Make a shallow-edged wooden frame to place on the existing roof, line it with plastic and fill it with soil. Plant it with grasses, sedums and wildflowers, or let the local plant population colonise it naturally.

~ **Wait before adding new plants.** If you've moved house and acquired a new garden recently, wait a while before you start digging up beds and borders. You'll avoid digging up and ditching treasures that have not yet appeared. Take pictures, make notes and record what appears and when.

   Look at the gardens around you for inspiration to see what grows well in your area.

~ **Think carefully about fence colours.** Bright shades are jolly but your tastes might change over time. Black, dark grey or green may seem oppressive when first painted, but once plants are positioned in front, the fence will blend into the background and the garden will seem larger, rather than smaller.

~ **Can you see it all in one go?** If that's the case, start off by creating a journey through your garden – even if it's tiny, it can still be divided up with small internal hedges or taller shrubs so you are tempted to explore and find hidden nooks that are lovely to sit in. Use trails of sand on the ground to mark out where you'd like beds. Leave them there for a few days while you eye them up and walk around, imagining what they'll be like.

~ **Make sure you put in your essentials.** You will need somewhere to sit, a path that is even – and interesting – to traverse, and screening from neighbours. But be considerate when it comes to fence height – 2m will give you privacy without robbing your neighbours of too much light.

~ **Make beds as deep and wide as you can.** Plant relatively closely in soil that has been enriched with organic matter. Not only will the plants grow better, they will also cover the ground faster and leave little room for weeds.

~ **Think vertically as well as horizontally.** Use arches, wigwams and pergolas to lift up your display wherever a tree would be too large or too dominant.

'Design your garden so it suits your lifestyle. If your time is limited, restrict the number of borders, or fill them with plants that need little maintenance.' **Alan Titchmarsh**

PROJECT – How to create a wall planter

When it comes to gardening in a small space every inch counts, but your planting space isn't limited to the square footage on the ground. Every garden, no matter how small, has some vertical space, be it wall or fence, that is ripe for populating with plants. A wall planter gives the option of growing a whole host of plants and will transform bare boundaries. They come in a range of forms, from expensive professional 'green walls' with irrigation systems to simple wall-mounted fabric or woven plastic pouches. These easy-to-install pouch systems allow even the most inexperienced gardener to nurture their own green wall and grow a huge range of plants, from vibrant bedding to flavoursome herbs. Try plants like ivy, succulents, campanula, *Polypodium vulgare* or aubretia. Look for plants that can cope with low levels of water, are compact and either trailing or rosette forming. Wall planters are also perfect for growing edibles near the kitchen door or brightening up patio walls, where there's no soil at the base.

1. Attach your planter to the wall, following the instructions for the type you have chosen. They can usually be attached with screws and rawl plugs. Fill your planter with compost. The type of compost you use will depend on which plants you choose, but aim for a peat-free multi-purpose. Fill each pocket of the planter around three quarters full with compost.

2. Settle the plants into their pouches, firming the compost around the roots as you go. Ensure the finished level of the compost is about 3cm below the ridge of each pouch. This makes for more efficient watering and less runoff.

3. With the wall planter now installed and the plants firmed in, it's time for a good soaking. Flooding the whole planter with water not only settles the compost but also ensures good soil–root contact so the plants can establish quickly.

4. Continue to water the planter through the growing season, ensuring the compost is moist 3cm down – check with your finger – and pay special attention during hot or windy periods. After the first six weeks, begin feeding with a general-purpose liquid feed once a month.

SPOTLIGHT ON
SUCCESSFUL BORDER DESIGN

Lack of planning can lead to borders that are difficult to look after or that have no cohesive design and look messy. It's daunting to be faced with a big patch of bare soil; your mind just goes blank. And it's always disastrous to start a border project with a trip to the garden centre. You'll come home with a maxed-out credit card and a car full of plants that you have no idea how to place – and half the time they don't even 'go' together. The mistake most people make – seasoned plantspeople as well as complete beginners – is to jump in with both feet first. A bit of planning goes a long way.

~ **Stick to simple outlines.** Tight curves look fussy and are hard to mow around.

~ **Try a theme** to give focus to a busy scheme – if you want to use a range of different plants, develop a theme such as grasses, herbs, cottagey or Mediterranean.

~ **Use one long-season, low-key plant repeatedly** through the border to pull together a display of bold, showy plants. Grasses are useful here.

~ **Include year-round interest.** Don't plan a border with only summer colour in mind. Draw up a series of storyboards to ensure your borders cover every season.

~ **Don't make borders too wide.** This makes them difficult to weed and can look out of proportion if your garden is small.

~ **Don't make them too narrow, either.** Narrow borders can look fussy and plants will flop onto paths or lawns.

~ **Try not to overplant.** This is a common mistake with plant lovers – it's worth checking the size plants will be after five or ten years in reference books or on plant labels. Even if you have researched well, some plants may outgrow their space. Be prepared to move some plants to safeguard frailer neighbouring plants.

~ **Dilute 'loud' borders** with neutral flower shades and coloured or variegated foliage, or for a subtle scheme restrict yourself to a limited colour palette.

~ **Add height with see-through plants.** Try light elegant species such as ornamental grasses, *Verbena bonariensis* and *Salvia uliginosa* which are leggy and have a small footprint, meaning they can be slotted in between the shorter plants in the middle of the border without taking up too much space. They add height without bulk.

~ **Provide year-round structure with shrubs.** There's a huge range of evergreen shrubs to choose from, with foliage shades from gold and burgundy through to silver. Variegated plants are worth considering, too.

~ **Avoid a messy border by repeating plants** through the scheme to bind it together. Buy several of the same plant, or large pots of perennials that you can divide.

3 top tips

- **In a small border,** take care to avoid 'thugs' that will spread rampantly and swamp everything else. Mint, Japanese anemones, bear's breeches (*Acanthus*) and the spreading reed canary grass (*Phalaris arundinacea*) can all take over a border.
- **Use neutral shades** such as olive green on your shed to make it less prominent and help it blend in with the planting. If you paint it a dark colour, such as black or deep blue, it will seem further away, helping to make a small garden feel bigger.
- **If you are designing your garden from scratch,** take a long, hard look at your plot and decide what you like about it and what you don't, what you could do without and what you'd like to add. Be realistic about your finances and the time you have available. There's no need to go at it like a bull at a gate. It'll take time. You're not decorating a room (which you can do in a weekend), you're starting on a magical journey that will continue to evolve.

GARDEN DESIGN CALENDAR

Garden design is not as dependant on the seasons as other gardening challenges, but many problems can be avoided by planning in advance, using the quiet months to install features such as sheds and storage boxes as well as order plants and make border plans.

Spring

~ Plant climbers to cover boundaries

~ If you've just moved in, wait to see which plants emerge so you can decide what you want to keep and what to get rid of

~ Spring is a good time to add new plants

Summer

~ Look out for gaps in your planting scheme and make notes on what to include next year

~ Plant up a summer container as a focal point on a patio

~ Look around gardens for inspiration

Autumn

~ Autumn is another good time to add new plants and plant trees and shrubs

Winter

~ Plan out any borders that need revamping or designing in the spring. This will give you time to work out which plants you need, and how many, and order them

~ Use the quieter months to fix wires to any fences that you will grow climbers up to solve the problem of bare boundaries

~ Assess your garden for winter structure

Q&A
COMMON QUESTIONS ABOUT GARDEN DESIGN

Which shrubs can be grown in a tight spot?

Don't imagine that the best way to make a shrub smaller is simply to chop off the ends of its branches. In many instances this will encourage the plant to grow back vigorously, proving the old adage that 'growth follows the knife'. You may well end up with a shrub that does not flower, looks butchered or just dies.

Choose your shrubs carefully, selecting those of modest ultimate size, or grow those that cope with being cut back hard each year, such as coloured-stemmed varieties of dogwood like *Cornus* 'Midwinter Fire'. They will show their brilliance the following winter when the leaves fall. But when treated like this, shrubs must be well fed and grown in earth that is slightly moist rather than dry. Alternatively, choose lower-growing shrubs such as potentillas, hydrangeas, cistus and evergreen *Rhododendron yakushimanum*, which seldom grows more than knee height and creates a spectacular show in spring.

What tree is best for a small space?

Even small gardens need height to lift them out of the ordinary, and trees are by no means out of the question if you choose the right kinds. Forget about weeping willows and other monsters that will cut out light and shift foundations, and instead seek out those of moderate growth and, if possible, more than one attractive feature that will offer interest for much of the year. The snowy mespilus (*Amelanchier*) has frothy white spring blossom followed by green leaves that turn orange and red before they fall in autumn. Crab apples such as 'John Downie' have wonderful blossom and attractive fruits that can be made into crab apple jelly, or left for the birds. Silver birch trees have light feathery canopies and white or cream-coloured bark that positively glows, even in the gloom of winter. Their leaves turn yellow in autumn. And in the smallest gardens, Japanese maples – varieties of *Acer palmatum* – are slow growing and wonderfully elegant, with leaves of green or purple that also offer brilliant autumn colour. Japanese maples can even be grown in large pots or tubs and will survive happily in such a situation for many years if the top 5cm of compost is removed and replaced with fresh each spring, and they are moved to a larger container every three or four years.

Which magnolia would create privacy in my small garden?

If you want privacy year-round, evergreen *Magnolia grandiflora* 'Kay Parris' has a dense, bushy habit, with small leaves, and flowers freely through the summer months. It can be pruned if it gets too large. It's one of the best evergreen magnolias!

Can you suggest trees for wildlife as a screen in my garden?

Have a look at those trees that give flowers in the spring followed by fruit and good colour in the autumn. Consider sorbus, malus and amelanchiers, but be sure to check their height and spread to select those suitable for your space – you don't want to plant anything that will grow too large.

Which easy plants can I grow for an attractive front garden?

Include a good mix of structural shrubs and perennials as well as some evergreen structure and repetition. Topiary balls or mounds – or maybe even more adventurous shapes – can take as little as two clippings a year, but really need no other interventions at all for the rest of the time. Yew (*Taxus baccata*) is a good option for this. To add a little looseness, how about some grasses, such as molinias or panicums? You can choose the right height to add the perfect amount of soft haze in the gaps between the planting. Or how about a good ground-cover plant, such as a hardy geranium or pachysandra, to unify the area and fill gaps?

How do I grow a succession of flowers in a small space?

To start the year, have lots of early bulbs like crocuses, narcissi and *Cyclamen coum*. I also love *Iris reticulata*. Then I would suggest species tulips, which don't need to be replanted every year. *Tulipa acuminata* has spidery red petals that reach up to the sky, while *Tulipa humilis* flowers early, in March or April, and is scented. You could follow these with fritillaries.

You could also sow a wildflower seed mix, but if your soil is fertile, it may encourage a single species to take over and out-compete the less vigorous ones. So instead you could opt for a perennial border using wilder varieties. Take a look at centaureas, as well as verbenas, pink *Pimpinella major* and various hardy geraniums. These should keep you going until the end of the summer.

What climbers can I grow in pots to hide a wall?

For containers, both shrubs and climbers will need as large a root area as possible if they are to thrive, so make sure the pots are a really good size – something like a half barrel or similar. They should have plenty of drainage holes. Fill them with a good-quality, loam-based potting compost.

When it comes to the plants, you can't have anything too big or that doesn't do well in a pot, and it must be manageable. Try *Rosa* 'Kew Gardens' with its open, single, white, repeat flowers, which grows to just over a metre and has a hint of perfume. You could have a few of these repeated down your wall in pots. Intersperse them with *Clematis* 'Diana's Delight', which has large blue flowers with white central markings. Both of these are attractive to pollinators. Or try a repeat of grasses in pots down your wall, such as a small miscanthus or calama-grostis. Alternatively, plant a fruit tree that does well in a pot, such as fig 'Brown Turkey' or apricot 'Moorpark' – both enjoy a south-facing aspect. All of these would have to be watered continuously on a sunny site.

Other options include annual climbers such as sweet peas, as they will bring rapid colour, perfume and screening, and they won't need such a large container either. Morning glory is another one that will put on a good show and won't get too big or too heavy in one season.

Planting in Problem Spots

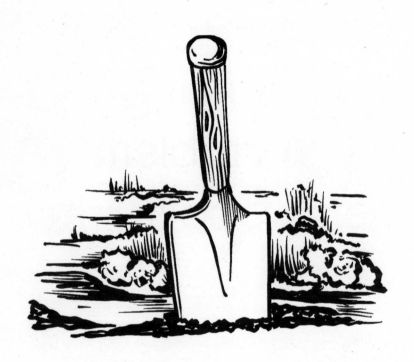

INTRODUCTION

Problem spots only become a problem when we attempt to place a plant where it doesn't belong. There are lots of amazing plants that love each different type of site, so a little research before you buy will save you not just money but disappointment too.

Understanding your soil is the first and most important step, and all you need is a simple soil testing kit from a garden centre. This will measure the pH, telling you if you have acid, neutral or alkaline soil. The next step is to get your hands into the soil and work out its make-up – is it sandy and dry, sticky clay, or silty?

Then look at which direction your garden faces (also called 'aspect'), and how much light each area gets during the day – does it bake in full sun for hours or stay mainly shady until late afternoon? Take time to understand how much sun your border gets – most borders are shaded for at least part of the day.

This simple information about soil and light levels should drive your plant choices. Instead of moaning about the conditions in our gardens, we should focus instead on the opportunities. Search online for information on the conditions specific plants need and read plant labels carefully before you buy.

FIVE COMMON
PROBLEM SPOTS

Shade

Shady gardens can be just as beautiful and rewarding as any sunny space. Just stick with the old saying 'right plant, right place', only choosing shade-lovers for a shady spot. You could create lush tapestries of foliage plants, punctuated with flowering bulbs in late winter and spring, and again in autumn.

There are different types of shade, and each will suit different groups of plants. Shady spots beneath trees are almost always dry, due to a combination of the umbrella effect of the tree canopy sheltering the earth from rain, and the double whammy of the tree's roots, which will suck moisture from the soil. Some areas are cast into deep gloom, perhaps thanks to the presence of towering, dense evergreens, while a deciduous shrub might cast more dappled shade, which is more amenable to shade lovers growing below. It's worth remembering that some plants are indeed shade-lovers, and others merely shade tolerant – up to a point. The depth of the shade – and how long in the day and in the season it lasts – will affect the plants that are able to grow there. Areas that are in shade for only part of the day – until the sun comes round and strikes them more directly – will grow a wider range of plants, including those that are only shade-tolerant, or suited to partial shade. Areas that remain in shade from dawn until dusk need plants for deep shade. Deep shade is a situation that most gardeners struggle with. Deep or full shade is usually classed as three hours or less of direct sun each day, and nearly all gardens have some, cast by trees, boundaries or buildings. The first thing to look at is whether the soil is dry or contains moisture. Is there clay or silt, or is it sandy and dry? There are lots of exciting plant combinations to try, especially those with interesting leaf shapes and textures. You could also try lightening shady areas by thinning tree canopies. Removing a few branches will allow more light into the understorey.

Clay soil

Clay soil does have some advantages – it's fertile and it retains moisture. This does mean that it can be sticky and hard to work in winter and baked solid in the summer, which makes it difficult for plants to grow. Heavy clay and ground where drainage is slow can be hugely improved by the addition of plenty of sharp grit. A combination of grit and organic enrichment will turn even the most intractable earth into something much easier to cultivate and much more conducive to plant grow than the 'plasticine' you started with. Plants that are good for clay soil include roses, foxgloves and daylilies.

Exposed and windy

Think of an exposed site, and coastal gardens come to mind. But inland gardens can be windy too, depending on their location and aspect, and whether there are buildings and boundaries nearby – either protecting them or creating turbulence. It can be difficult to establish plants on a windy site, so when you're planting you may need to provide temporary shelter. Look at other local gardens or the wider landscape for planting inspiration. Plants that can cope in these sites include eryngiums, tamarisk (a shrub or small tree that also makes a good windbreak) and centranthus, which is also happy in poor, well-drained soil.

Damp

If you notice that ferns, moss and mind-your-own-business are thriving in your garden, you may have wet soil. Holes may fill with water when you dig and plants can be prone to fungal diseases. Waterlogged soil can be tough to cope with in winter, but in summer it has great potential to come alive with colour, just when other areas are becoming more difficult. Many plants hate having their roots in water, so instead of making them struggle, embrace your wet soil and create a bog garden full of vibrant plants that love these conditions such as Siberian irises, rodgersias and trollius. Rodgersias are fantastic for bog gardens, with big textured leaves that are really dramatic. If you really want to grow plants that prefer a less soggy site, you'll need to improve the drainage by digging in lots of organic matter and coarse grit or installing drains.

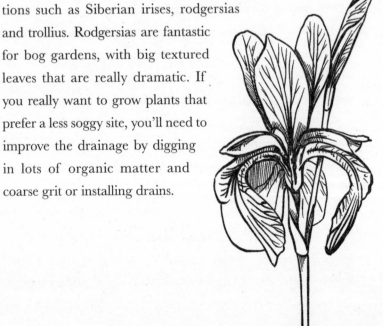

Dry

Soil that is either sandy or very stony tends to drain really quickly. It's easy to dig and warms up quickly in spring, but nutrients are easily washed out, as is any moisture. Dry shade is probably the most difficult to work with, so dig in plenty of organic matter every year and keep the planting simple. Mulch the soil when it's damp to retain moisture, with either compost, bark or gravel. Try a few different plants, and find two or three that work well and then repeat them. Mediterranean plants will enjoy the good winter drainage and other plants to try include sedums, thymes and artemisias. Add generous amounts of well-rotted manure or garden compost, which will aid moisture retention in dry soils.

'There is no doubt about it, shade worries gardeners. It is seen as a problem spot, a dreary spot and a spot where nothing will want to grow. Well, nothing could be further from the truth. There are literally hundreds of plants that will thrive in shade, because that is where they choose to grow in nature. It is up to us to seek them out and make use of their abilities, rather than try to force sun-lovers to grow where they will struggle to survive.'

Alan Titchmarsh

ACTION PLAN
HOW TO IMPROVE
GROWING CONDITIONS

~ **Add organic matter to your soil.** Regularly add lots of organic matter to your soil, such as garden compost or well-rotted manure. This coats the mineral particles in the soil, improving the drainage of heavy soil and the nutrient- and water-holding ability of light soil.

~ **Check your soil's texture.** Work out what type of soil you have so you can buy plants that will be happy in those conditions. If you have moist soil you should be able to squeeze and mould it. Clay and silt stick together, sand and chalk don't.

~ **Mulch regularly.** Mulching the surface of the soil helps to hold in water and reduce evaporation. Apply mulch in a thick layer to damp soil. Garden compost, pebbles, bark or wood chips are all suitable choices. They will also suppress weed growth, and garden compost will have the added bonus of feeding the soil too.

~ **Right plant, right place.** Choose plants that are natu- rally adapted to low light levels when looking for plants for shade. This is the most important factor for successful shade gardening.

~ **Water new plants,** even if plants are specifically for dry shade or drought conditions – just because a plant can tolerate dry shade it does not mean that it will do so from the word 'go'. It will need watering regularly in its first season to encourage the roots to establish themselves and spread out in search of moisture, so that in future seasons they are better equipped to sustain the plant.

*'I want my plants to be happy and healthy,
so I take my lead from nature and grow things
that naturally enjoy the conditions I have.'*
Adam Frost

PROJECT – How to test your soil

Testing the soil in our gardens is vital if we want to know which plants will do well and which ones will struggle, thus avoiding wasting money. Check how heavy or light your soil is as well as its pH.

Have a dig around to see what type of soil you are dealing with. If your soil is gritty and not sticky, it's likely a free-draining sandy soil and Mediterranean plants are the order of the day. If your soil is heavy and very sticky when wet, then it's likely to be clay, and roses and hardy geraniums will thrive. Many soils will be somewhere in between, which allows you to grow a wide range of plants.

On very alkaline soils (which overlie chalk or limestone), some plants are unable to extract iron from the soil, which results in their foliage turning yellow or chlorotic. The technical term for them is 'calcifuges' – plants that need acidic soil to do well. Other plants thrive in alkaline soils and have no trouble extracting nutrients. These plants are known as 'calcicoles'. Soil acidity and alkalinity is measured on the pH scale, which ranges from 0 (extremely acid) to 14 (extremely alkaline). Most soils in the UK have a pH of between 4.0 and 8.0, and pH 7.0 is neutral. Testing your soil will guide you to the plants you can grow next year, though looking at what thrives in neighbouring gardens is also a good indicator. A pH testing kit will give a good indication of the soil acidity or alkalinity, helping you choose the right plants for the soil.

1. Collect small soil samples from the part of your garden you wish to test. Bear in mind that the pH may vary slightly across the garden (though seldom extremely).

2. Dry the soil near a radiator or other heat source and remove any stones or debris that could skew the reading. Mix up the soil samples and add to a test tube.

3. Pour the indicator fluid in the quantity suggested – usually up to a line marked on the test tube. This fluid is specially formulated to detect the soil's acidity or alkalinity.

4. Shake the soil and fluid for half a minute, then let the mixture settle to reveal the colour of the fluid. Check this against the chart to find out its pH value.

SPOTLIGHT ON
DROUGHT CONDITIONS

As temperatures rise, many of us have watched our plants struggle in the summer heat and wondered how best to help. As gardeners, we want solutions that won't impact too much on the environment, which is why using water is such a dilemma. On the one hand we hate seeing our prized plants hanging on for dear life, but on the other we want to do our bit to save this precious resource whenever we can.

We do, however, need our plants to be strong and healthy, and a plant that is weakened by drought is more susceptible to diseases. Also, there are some things you really do have to water: a freshly laid lawn, newly planted specimens (especially trees and shrubs, but also perennials), anything in containers, and annual vegetables.

~ **Aim water at the soil around the base of plants** rather than at the plant itself, as they take up water through their roots. This will also help to avoid leaf scorch, which can occur when hot sun shines on damp foliage.

~ **Water in the morning or evening** once the sun has lost its heat. Evaporation occurs when liquid water turns into a gas – water vapour. If you water in hot sunshine, the moisture will evaporate from the soil before your plants have the chance to take it up.

~ **Time your planting well.** If you lay turf and plant trees, shrubs and herbaceous perennials in spring or autumn, then nature waters them for you. Consequently they need far less additional watering while they get established.

~ **Consider drought-tolerant options** when adding new plants. These often have silvery leaves, such as lavender, artemisia and stachys. Other plants to try include phlomis and red hot pokers, which love a sunny spot.

~ **Water enough to benefit the plant every time you water.** Just sprinkling the soil lightly encourages plants to grow roots near the surface – an area that dries out readily. Watering into a buried tube sends the moisture deep into the ground and encourages long roots that are better able to sustain the plant. Once a plant is established, you should let it find its own water.

~ **Mulch the surface of the soil** to help hold in water and reduce evaporation. Apply mulch in a thick layer to damp soil. Garden compost, pebbles, bark or wood chips are all suitable choices. They will also suppress weed growth, and garden compost will feed the soil too.

'Plants manage to survive in the wild with absolutely no input from us and they can do the same in your garden. The worst thing you can do is mollycoddle them so much that they can no longer cope on their own. Roots do a great job of going down into the soil to find the water table. There is a balance to be struck though – we can't afford to replace dead plants every summer and we don't want to just stand by and watch them die, especially when we could have saved them. There's nothing wrong with watering (using only a can if there's a hosepipe ban) if you do it carefully to avoid wastage.'

Frances Tophill

3 top tips

- **Improve a windy plot.** Plan to plant some tough shrubs in autumn, on the side the wind comes from. In the meantime, you can make a wind break with woven fabric and sturdy posts. Include a few gaps so that a little wind can pass through, rather than blowing the windbreak down.

- **Cut back plants that are prone to wind rock,** such as shrub roses, lavatera and buddleia, by a third to half of their height in autumn and bring vulnerable ones like olive trees indoors if you can.

- **Illuminate a shady spot with leaves.** Brighter is better when it comes to foliage. Vibrant and variegated leaves help to illuminate the garden, yellow and golden highlights add warmth, while leaves with cream edges or markings are the most reflective, bringing a shimmer to sunless corners. Use evergreens, such as *Euonymus fortunei* 'Emerald Gaiety', for year-round light.

DEALING WITH PROBLEM SPOTS CALENDAR

Seasonal care for problem spots includes soil improvement, planting at the right time to establish plants and getting to know your plot.

Spring

~ Spread a thick layer of mulch on your borders to improve the soil

~ Order or buy new plants for gaps, making sure to check which conditions they will thrive in

Summer

~ Water new plants, even drought-tolerant ones, until they are established

Autumn

~ Plant shade-tolerant, spring-flowering bulbs to brighten gloomy spots

~ Mulch borders to improve the structure of clay and dry soils

~ Make leafmould as a soil improver

~ This is a good time to plant a hedge or add shrubs and trees that make good wind breaks

~ Prune any shrubs or trees at risk of wind rock

Winter

~ Test your soil before the growing season begins
~ To reduce soil acidity, add lime in winter every few years. Spread garden lime or calcified seaweed on the soil and fork or rake it into the surface
~ Tie in climbers in exposed sites to avoid wind damage
~ Thin out the canopy of overgrown trees to let more light into shady borders

'As with all gardening, if you choose plants that originate in conditions similar to those they are likely to experience in your garden, they should thrive and flourish.'
Carol Klein

Q&A
COMMON QUESTIONS
ABOUT PROBLEM SPOTS

What will grow in shade apart from weeds?

Many beautiful woodlanders love shade. They include foliage plants such as bergenias, brunneras, hostas, Solomon's seal, ruscus, lily-of-the-valley, actaea, daphnes, aspidistras, euonymus and all manner of ferns. Choose a mixture of evergreen and deciduous plants. They'll look best in groups of three, five or seven, rather than singly. Repeat the groups too if you have lots of space to fill.

Why does everything I plant under my tree die?

Perhaps the hardest part of gardening in shade is establishing plants under mature trees. The soil can be very dry and poor, so plants need a lot of help to settle in. Lay a seep hose (a special hose with holes) through the bed, coiled around the plants, to water easily and effectively. Then every winter add a thick layer of mulch (garden compost or well-rotted manure) to seal in the moisture before spring. Also consider removing some of the tree's lower (but not the top) branches to reduce the depth of shade.

Can I have flowers in my shady garden?

Many bulbs are happy in shady spots. If your shade is cast by deciduous trees, that's ideal, as late-winter and early-spring bulbs grow up and flower before most trees come into leaf. You could start with bright yellow winter aconites (*Eranthis hyemalis*), snowdrops (*Galanthus*) and a lovely white daffodil (such as *Narcissus* 'Thalia') to give a visual lift. Then add colchicums for autumn flower power. Both spring- and autumn-flowering cyclamen are great choices for shade too.

What is horticultural grit made from?

Normally it is either made from crushed granite or limestone and is crushed to be 1–4mm in size. The granite is acidic in nature, the limestone grit is alkaline, and they should be used accordingly. Neutral pH grit is also available. The particles should be angular not round. It improves aeration in heavy soils and may be used for insulation.

How do I plant an area that's both in sun and shade at different times of the year?

Improve the soil with well-rotted organic matter so it does not dry out completely, then plant spring-flowering bulbs, which go dormant through summer. Try narcissus, wood anemones, trilliums and corydalis. Some ferns also tolerate these contrasting conditions, including forms of our native evergreen *Asplenium scolopendrium*, such as Crispum Group, and *Dryopteris erythrosora*. The more sun-loving hostas will flourish in this situation. This type of shade, where a building, for example, blocks sun out for part of the year, means that it's best to avoid sun worshippers like silver-foliaged Mediterranean and marginally hardy plants, which will find such a spot difficult during cold, wet winters. Heavy shade can also make many plants start into growth a tad late. However, after the frosts, you could plant out tender bedding as the sun will be available for most of its flowering period.

What shrub can I plant on a west-facing slope?

West-facing sites get a little shade in the morning followed by afternoon sun. This means it's best to stick to plants that like full sun or partial shade. For something unusual, try a Japanese plum yew (*Cephalotaxus harringtonia* 'Fastigiata'), which has leaves like a yew and edible fruits that resemble plums. It's evergreen, can deal with sun or part shade and is very hardy to -20°C. It grows to about 1m wide and 2m tall. A more familiar and readily available option is an Oregon grape (*Mahonia aquifolium*). It has glossy evergreen leaves and clusters of bright yellow flowers in spring. It generally prefers full or part shade, but can tolerate sun if the soil doesn't dry out. The flowers are an excellent early food source for pollinators, and are followed by dark berries for birds.

What can I grow on a south-east facing terrace?

With this sunny aspect, there is a lot you can do in a small space. Use the largest containers you can find/fit/afford as they will always yield the best results, be less likely to blow over and need less watering. You could grow your own veg, with beetroot, courgettes, dwarf French beans, summer salad crops and tomatoes. You'd also be able to house a neat collection of herbs, with trailing rosemary in hanging baskets to make the most of your vertical space.

If you want a purely ornamental display, especially if you have a modern-style apartment, you might like to try something contemporary with palms and succulents, using lots of blues and silvers from plants like the hardy blue fan palm, *Chamaerops humilis* var. *cerifera*, with pots of *Echeveria elegans*, and the ever-reliable and stylish *Agave parryi*. Add height with the silvery foliage of *Cupressus arizonica* var. *glabra* 'Blue Ice'.

Which colourful evergreens will suit a small, shady garden?

Variegated evergreens will provide year-round leaf colour and structure, and you can add ground cover and bulbs beneath. The compact shrub *Skimmia japonica* 'Perosa' has green leaves with cream margins and produces bright-red buds and cream flowers in winter. The foliage of *Aucuba japonica* 'Picturata' has a bold splash of yellow and yellow-blotched edges. If your soil pH is acidic then try compact *Rhododendron yakushimanum* 'Koichiro Wada'. Its new leaves are silvery white, turning deep green with brown undersides. Its pink flower buds open into white blossom-like clusters in May. If your soil is unsuitable, grow it in a container. Do check your soil before buying shrubs, to make sure they will suit your conditions. Many camellias do well in shade and there are lots to choose from. Daphnes are renowned for their highly scented pink or white flowers. Some varieties bloom as early as February, while others flower from early spring through to summer.

Ornamentals

INTRODUCTION

Try as we might, there are always things that don't go according to plan – areas of the garden that aren't performing as we had hoped. But with a little ingenuity and forward planning, many of these irritations can be solved. Crack the problem and your gardening will be more enjoyable as a result. Pests and diseases are responsible for many of the problems that come up when growing ornamental plants, but there are also problems caused by weather as well as getting your timing or planting techniques wrong. Luckily there are solutions and ways to prevent many of these problems, whether it's frost protection or mastering a few easy skills that will ensure seeds germinate and plants thrive.

FIVE COMMON PROBLEMS WITH ORNAMENTALS

Plants stopped flowering

Many plants, such as sweet peas, will stop flowering if they are allowed to run to seed. Keep picking flowers to bring indoors and deadhead regularly to keep plants flowering right through the summer and into autumn. Not only is this technique useful for prolonging flowering, but the garden will also look a darned sight better for being free of faded flowers. Verbena flowers for a long time even without deadheading, but you may find it spreads its seed a little too freely. So, remove faded flowers to keep it looking neat and stop it popping up where it's not wanted. Other reasons for a lack of flowers could include poor pruning (at the wrong time of year), planting position or the plant may not yet be established enough to flower well.

Bulb blindness

Daffodils with bulb blindness have plenty of healthy-looking foliage, but few flowers. Either no buds are formed, or there are a few, small, empty buds. Blindness is most common on double and multi-headed varieties, and on daffodils that have grown in the same place for several years. If lifted, the bulb itself appears perfectly normal. Blindness suggests the bulbs have literally run out of energy and do not have sufficient resources to produce a full complement of petals; indeed sometimes none are produced at all. Lift, divide and replant congested old clumps of bulbs in the autumn, feed them regularly and within a year or two they should be flowering again. Blindness can often be prevented by watering regularly during dry summers, as plants can't take up the necessary nutrients to grow when the soil is too dry. Another way to limit the problem is to apply a foliar feed while the leaves are green, along with a slow-release fertiliser.

Seeds won't germinate

It's frustrating when you go to the effort of sowing seed and then nothing happens. To germinate, seeds have three basic needs: moisture, correct temperature and oxygen. They also need to be sown at the correct time. Check the seed packet for instructions on which month to sow. Some seeds need a high temperature to germinate, while others need special techniques such as soaking, scarifying (nicking the seed coat) or a period of cold. For beginners, start with seed that doesn't require advanced techniques or take weeks to germinate. Finally, make sure your seed is in good condition: seed viability can be affected if it is old or has been incorrectly stored and allowed to get wet or mouldy.

Keeping plants alive while away

When you're on holiday, the last thing you need whirring around in the back of your mind is the worry of returning home to dead plants and a wreck of a garden. Most of the potential problems revolve around drought or the fact that plants keep growing when you're not there, but a few simple tasks will get your garden prepared.

Move pots to a shadier area and stand them in saucers to collect water or set up an automatic irrigation system with drip lines. Avoid planting new plants before going away as they will struggle to establish if the ground is dry, but if that's unavoidable make sure you water well around the plants and apply an organic mulch about 10cm deep to help seal in the moisture. Any pot-bound plants will dry out quickly, so try to pot these on before you go.

Before you leave, deadhead and clear away flowers, including any that are just about to open. These will only flower and waste energy setting seed in your absence if left, so it is best to be ruthless before going away. If it looks like the weather is going to be very hot, consider setting up an automatic watering system for the plants that need regular watering.

Frost

As temperatures drop, frost can lead to leaf and stem damage, heaving of the plant crowns (when cracks in the ground expose roots to air) or even plant death. But with the right plants and the correct protection it's possible to deal with another beastly blow from the east relatively unscathed.

Protecting plants from damage is relatively straightforward. Below ground plants such as dahlias and bulbs can be lifted out of the ground over winter or protected with a deep, dry mulch. Vulnerable evergreens and shrubs can be wrapped in fleece (with good ventilation) or boxed into homemade polycarbonate frames. Tender plants should be lifted before the first frost and only planted out after the last predicted frost of spring. Container plants are best placed in a sheltered spot, while tender perennials such as penstemon are protected from frost by their own dead stems if you avoid pruning them back until spring. It's worth considering the situation of plants. Early flowers such as camellia and magnolia are often damaged as a result of thawing too quickly after an air frost. The simple solution is to grow them against a west wall.

'Though many plants have evolved to cope with the chill, others grown in Britain hail from such warm climes that they can't survive without our intervention. Over the years I've experimented with myriad ways of protecting plants. A simple mulch of dry material such as straw or bark chips works well to shroud the dormant tubers of tender species such as dahlias and cannas. I wait for the frost to "blacken" the plants before cutting them back and applying a layer of mulch for warmth, with plastic over the top to protect from rain.' **Nick Bailey**

ACTION PLAN
HOW TO GET THE BEST FROM PERENNIALS

~ **Plant at the right time.** Plant container-grown perennials at any time of year, but avoid planting during frosty weather, in spells of prolonged drought (the plants will fry unless you are a slave to the watering can) or in winter in heavy clay soil (the plants will drown). Plant bare-root perennials (and that includes your own divisions) in the autumn or spring months.

~ **Add organic matter.** Enrich the earth with a generous supply of organic matter (well-rotted compost or manure) before you plant, give the ground a good dusting of blood, fish and bonemeal for good measure, and remove all perennial weed roots.

~ **Plant well.** Dig a hole just large enough for each plant's roots and firm the earth back, making sure that the crown of the plant (where roots meet shoots) is level with the surface of the soil. Do not plant too deeply. Water the plants well, then spread a 5cm mulch of chipped bark, compost or manure on the surface of the soil, taking care not to swamp the plants themselves. This will hold in moisture and keep down annual weeds.

~ **Care for your plants.** Stake taller varieties with twiggy pea sticks or wire plant supports before they need it. There is nothing worse than a trussed-up perennial that has been staked too late in the season. Firm back the earth around the plants after frost and do not allow them to go short of water in their first summer.

~ **Cut back in autumn.** Cut back herbaceous perennials right to ground level in late autumn. Do not leave an inch or two of stalk behind, which will cut your fingers when you come to work with them the following spring. Compost the faded stalks.

Project – How to divide overgrown clumps

If clump-forming perennials get too big, dividing them will help keep your plants healthy and growing well. Established perennials can be dug up and divided in autumn or spring. In the south of the country, and on light, free-draining soils, it pays to do this in autumn so that the new plants can establish their roots and grow away vigorously the following spring. On heavy clay soils, and in the north of the country, spring planting is safer so that the plants do not have to spend a winter in cold, wet earth, which could lead to rotting. That said, spring planting is safe wherever you live, and provided the plants do not go short of water they will establish rapidly. This works best for plants with fibrous roots that are easy to dig up and divide, such as hardy geraniums, hemerocallis (daylily), astrantia and tradescantia.

1. Use a spade (or a fork if your garden's ground is especially hard) to dig up an established clump. Dig around it, prising the plant from the ground when you have cut all around it.
2. Chop the clump into sections with a spade.
3. Discard the central section of the plant if it is old and tired, then chop up the healthy, vigorous parts of the clump into pieces as large as your fist, or larger if you have enough for your needs.
4. Plant the new divisions wherever you want to place your new plants, and water well.

SPOTLIGHT ON
SUCCESSFUL SEED SOWING

If you've always been daunted by the prospect of sowing seeds, make this the year when you have a go. Yes, things can go wrong, but seldom if you bear in mind a few basic guidelines. Follow the instructions here and on the packet (always worth a glance!) and you'll find that the thrill of raising plants this way is far and away greater than buying them in pots, and a darned sight less expensive, too.

~ **Don't let the compost dry out.** Once water has been applied to freshly sown seeds, the germination process has begun – it's irreversible and desiccation at that stage means death.

~ **Try soaking hard seeds.** Sweet peas and other hard-coated seeds may germinate faster if soaked in water overnight before sowing.

~ **Never sow more deeply than twice the diameter of the seed.** If seedlings fail to appear, the seeds may have been sown too deeply or may require light to germinate.

~ **Always use fresh seeds.** Seeds lose their viability with age – sometimes after only a year.

~ **Don't leave seedlings too long before pricking them out.** They get overcrowded. This makes them weak and spindly, stretching for the light. Like this they're prone to damping off, a fungal disease that can cause them to keel over.

~ **Sow seed at the right time.** Seeds sown too early may suffer due to low temperatures (if not given enough heat) and fail to thrive. Seed packets give recommended temperatures for sowing and growing on.

~ **Use peat-free seed-sowing compost.** It is relatively inexpensive and will give you every chance of success. Homemade compost from the compost bin is not suitable for sowing seeds – it may contain impurities that harm them, so is not worth the risk.

'There is something magical about sowing seeds. After all, how can anything so green, vibrant and full of energy grow from something so small and dry and seemingly devoid of life? But with the addition of water, air and a suitable temperature, the most elaborate and majestic plants can emerge from a pot or tray of compost into which seeds have been sown.' **Alan Titchmarsh**

3 top tips

- **A good rule of thumb** is to plant any spring-flowering bulb at least three times as deep as the height of the bulb, or deeper. This means that a bulb measuring 7–8cm from base to tip needs to be planted with about 20cm of soil covering it – so the hole should be 27–28cm deep.
- **Check tender shrubs** and evergreen perennials for 'frost burn' (scorched leaves) and cut out damaged parts once the weather has warmed. Note the affected plants so you can protect them with fleece next winter.
- **Thin out some of the stems** of overcrowded perennials in spring to encourage strong, compact growth – a good technique for plants like asters and sedums (now also called symphyotrichum and hylotelephium). Cut out a third of the young, leafy shoots to give the remaining shoots plenty of light and more space for flowerheads.

GROWING ORNAMENTALS CALENDAR

Help prevent problems with ornamentals by keeping up a good care routine throughout the seasons.

Spring

~ Lift and divide perennials that have got too big

~ Mulch borders with organic matter

~ Protect plants with fleece if frost is forecast

~ Thin out some of the stems of overcrowded perennials

~ Sow annuals successively for a continual show of flowers through summer

Summer

~ Put in supports for dahlias to prevent the stems snapping under the weight of the flowers later in summer

~ Remove rose suckers before they take over

~ Feed plants to keep them healthy and encourage a good display of flowers and foliage

~ Plant out tender flowers

~ Move containers into the shade if you're going away

~ Water new plants regularly in dry weather

~ Deadhead regularly to keep flowers coming

Autumn

~ Shorten the stems of any shrubs, such as overgrown shrub roses, that could be prone to wind rock – these can be pruned properly in spring

~ Overwinter tender plants

~ Plant spring-flowering bulbs, making sure they are planted deep enough. Don't plant tulip bulbs until November to prevent the risk of tulip fire disease

~ Protect vulnerable plants from the cold

Winter

~ Tie in climbers to prevent wind damage

~ Cut back perennials to make way for emerging new shoots

Q&A
COMMON QUESTIONS ABOUT GROWING ORNAMENTALS

My once-healthy tulips have blotched and distorted leaves. What's gone wrong?

The numerous tiny greyish or bleached spots on tulip foliage and petals are caused by the fungus *Botrytis tulipae*, the cause of tulip fire. In extreme cases the whole plant is distorted and discoloured, or may fail to appear above ground. Remove infected foliage and flowers, or entire bulbs if necessary, discarding any that have tiny black spores (fungal resting bodies) on them. This will help limit spread. Plant healthy tulip bulbs in a fresh site and in future move your tulips to a different spot each year, ideally not using the same site for at least three years. Late planting, from November onwards, will also help.

Can you tell me why my agapanthus never blooms? I get plenty of foliage but no flowers.

This could be due to overfeeding, either by planting it in good soil, or being too generous with food and water if it is in a container. Agapanthus will flower best if it has restricted roots and is given limited feed.

Why won't my tulips re-flower?

Tulips are notoriously tricky. Deep planting is essential, but they also need a sunny spot and free-draining soil to stand a chance of flowering year after year. If you want to grow them in shade or heavy soil, it's best to plant tulip bulbs in deep pots that can be plunged into the ground while they're in flower then moved to a sunny, sheltered but less conspicuous spot where they can die down naturally. A good rule of thumb is to plant any spring-flowering bulb at least three times as deep as the height of the bulb. It won't hurt to plant deeper than this, particularly on light, free-draining soil that is likely to dry out more in the summer. If you want to get the best possible display from your bulbs, you must leave the foliage in place for at least six weeks after flowering. Even better, leave it to die down naturally without cutting it off. Plants will also benefit from a scattering of sulphate of potash to encourage the development of flower buds.

Why did my seedlings collapse?

This is almost certainly due to post-emergent 'damping off', caused by fungal diseases phytophthora pythium and rhizoc-tonia. The symptoms are collapsed seedlings with thin, brown stems, sometimes with a fluffy white coating. Cool, humid conditions and overwatering can make the problem worse and increase the chance of disease taking hold, as can sowing seeds too thickly, which stops air circulating between the seedlings. To prevent damping off, start by making sure everything you use is clean – compost, containers and water. If you make your own potting compost, use an electric steriliser to kill any fungal spores. Sow seeds thinly and remove any damaged seedlings.

Why does my thriving mahonia never flower?

It may be too young. Some larger varieties do take time to settle and produce strong, mature flowering stems. Mahonias shouldn't be pruned during this time, but if it's essential, only thin out a few older stems (usually after flowering), removing them down low.

Although mahonias are well known for shade tolerance, deep shade with no direct sun at all could prevent flowering. Consider pruning back any nearby trees and shrubs to let in more light. It's also possible your mahonia is doing too well, so if you've been watering and feeding it, ease back and leave things to nature. In particular, steer clear of nitrogen-rich fertilisers that promote leafy growth. But you could consider a high-potash feed to help stems mature, ready for flowering.

Why hasn't my wisteria flowered?

This is a very common problem for wisteria growers. Remember that plants are unlikely to produce flowers before they are established, which can take up to five years. They also need a sunny position. Too much nitrogen fertiliser can result in more foliage than flowers so try a high-potash feed to promote flowers. The biggest reason for a lack of flowers is poor pruning. Wisteria needs to be pruned twice a year. Cut back the current season's growth to five or six leaves after flowering in July or August and then to two or three buds in January or February. This will encourage short flowering spurs to form on your plant.

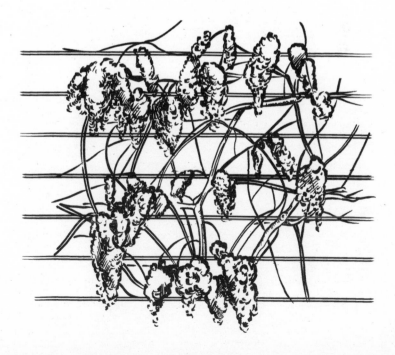

Fruit

INTRODUCTION

Fruit growing can seem like an expert's game: all that pruning to maintain shape and encourage regular cropping; those mysterious rootstocks with their strange letters and numbers, onto which the cultivated varieties are grafted as well as the regular pests and diseases. It all seems fraught with peril. But fruit growing can also be a low maintenance way to grow your own crops. Once plants are established and you are used to a pruning routine, your plants will produce crops year after year without the need for sowing and potting on. The key to tackling problems such as pests, diseases or disappointing crops is recognising what you're dealing with so you can take the appropriate action. Start off slowly with easy to grow fruits like strawberries and raspberries and you'll soon find it rewarding rather than daunting.

'The easiest things to grow? Raspberries and strawberries without a doubt. "Soft fruits" we call them, as distinct from "top fruit" which grows on trees. Strawberries demand nothing more than a decent patch of well-drained earth in full sun. Raspberries would be the last crop I'd give up for they are amazingly obliging.'
Alan Titchmarsh

FIVE COMMON
FRUIT PROBLEMS

Wasp damage

Thin-skinned fruits such as grapes, plums, peaches and strawberries may be damaged directly by wasps biting and chewing them. On apples and pears and other tough-skinned fruits, the fruit is generally injured first (by a scab attack or bird peck perhaps), and the wasps then feed on the wounded area. It may be worth protecting individual fruits or trusses of fruit with fine net bags (alternatively, use old tights or stockings!) or by making wasp traps using a jam jar part filled with a sugar and jam mix.

Flowers but no fruit

The most likely cause for this is frost damage. Frost can kill flowers that are produced early in the year and it also discourages pollinating insects. Fruit trees that produce flowers early include plums and cherries. Look out for bad weather and cover your trees with fleece if a hard frost is forecast. The other reason could be a lack of pollination – perhaps the tree is too far from pollinating partners. Fruit such as plums and apples benefit from being near other trees. Look for self-fertile varieties and encourage pollinating insects into your garden.

Common scab

Rough, sometimes corky, khaki grey-coloured scabby patches develop on the skin of apples and pears, sometimes resulting in fruit cracking, or distortion. Severe infections often follow damp weather but, unless the fruit splits, it is generally perfectly good to eat once peeled. Rake up leaves under infected trees and prune out scabby shoots when winter pruning. Don't let the centre of the tree get overcrowded as this prevents air circulation which can lead to the humid conditions that scab thrives in. If your tree is particularly susceptible, try to replace it with a scab-resistant variety.

Winter moth

Caterpillars eat irregular holes in leaves and produce silken threads that bind leaves together, forming a safe 'pouch' in which to hide. They also eat fruit tree blossom and may attack fruitlets, causing deep clefts. The yellow-green caterpillars are 2.5cm long, with pale lines running along their length, and they walk with a 'looping' action, arching their bodies with each forward move. Adult winter moths appear in late autumn and winter, emerging from pupae in the soil. The wingless females climb up into trees to mate with the winged males and lay eggs on the foliage. The eggs hatch in spring, as soon as the temperature reaches 13°C. Attach sticky bands (grease bands) around the trunks of trees in the autumn to prevent the females from climbing. Make sure you keep the bands free from debris that can act as a bridge, allowing females to cross the band. Entice birds into your garden with feeders and nest boxes as they'll feed on winter moth caterpillars in spring.

Brown rot

If the fruit on your trees turns brown it could be brown rot. This happens when fungal spores get into the fruit through wounds in the tree, either from pruning or damage from insects and birds. It affects apples, pears, plums and other tree fruits such as cherries and apricots. Dispose of any fallen fruit and any brown fruits on the tree. Cut back a section of the stem to just above a leaf to remove the infection. Thin fruits in early summer to improve air circulation around them.

'In the past, domestic fruit growers would have a complex spraying programme against pests and diseases. I see little point in smothering something I'm going to eat in noxious chemicals, so I use none of them. Yes, it means that I have to cope with a bit of apple scab, but that can be removed by peeling the fruit, and the odd maggot will rear its head, but the majority of the fruits are healthy and tasty, thanks to good growing conditions and a mixture of crops rather than a concentration of one type in any one area.'

Alan Titchmarsh

ACTION PLAN
HOW TO SUCCEED WITH FRUIT

Do your research

Read up about rootstocks before you buy to make sure you have the right one for your soil and space. Look for types of fruit that will suit your space. Cordon or stepover apples and espalier fruit are a good solution for small gardens.

Get the planting right

Plant bare-root fruit between November and March, but firm back any that are later loosened by winter frosts. Container-grown trees and bushes can be planted at any time of year but avoid planting during periods of drought. Never plant on waterlogged ground and make sure the soil has good drainage. Improve the soil before planting with a generous helping of organic matter. Improve sticky soil by adding sharp sand and grit.

~ In grass, maintain a bare area of soil 1m wide around fruit trees, to reduce competition and make feeding easier.

~ In pots, make drainage a priority. Use peat-free John Innes No.3 compost for pots rather than a lighter, soilless multi-purpose mix, which won't have the weight or sustenance needed for most fruits. Use an ericaceous compost for blueberries.

Choose the right site

Choose an open aspect so the sun ripens stems and fruit. Avoid spots that are exposed to strong winds or low-lying areas that are frost pockets, which will lead to blossom damage and crop failure.

Maintain good aftercare

Dwarfing rootstocks will need staking for most of their life. Check ties regularly to avoid swelling stems being constricted. Don't let plants dry out completely, and feed them regularly.

PROJECT – How to plant bare-root strawberry runners

Every garden has room for a few strawberries, and everyone enjoys picking and eating these tasty fruits. A small patch of well-drained earth in full sun is all they need to grow and crop well. For a successful crop, beef up the soil with garden compost or well-rotted manure, as the plants will be there for at least three years, then add some blood, fish and bonemeal. To reduce the need for weeding and help retain moisture, you can plant through a membrane. This means you won't need to lay straw around the plants at fruiting time. Remember that strawberries will need to be replaced every three years.

1. Unpack the bare-root plants as soon as they arrive and soak the roots in water for at least an hour and preferably overnight.
2. Weed the bed thoroughly, then cover with membrane. Cut crosses in the membrane at 30cm intervals where you're going to plant. Hold it back with a trowel to plant.
3. Set the plant in the hole so the point where the roots meet the leaves is at ground level. Firm back the earth around the roots, then fold the membrane down to cover the soil.
4. Water well to settle the soil round the roots, then water regularly during any dry spells. The membrane will help to retain moisture.

SPOTLIGHT ON
GETTING A
GOOD HARVEST

~ **Pollination is vital.** A wide range of plants flowering at differing times in your garden will attract a variety of pollinators, so choose different species and cultivars. If there aren't enough pollinators around, dab a paintbrush from flower to flower to guarantee pollination. This will be a must if you're growing greenhouse citrus or peaches.

~ **Apply a regular high-potash feed.** As well as a mulch of compost or manure in winter, use a liquid feed, such as seaweed or tomato feed, throughout summer to promote flowering, fruit formation and fruit ripening. Container fruits like citrus need top dressing with fresh compost every year or two, as well as regular feeding when fruiting.

~ **Water at the right time.** Without regular water the fruits will shrivel, then split after heavy rain. Consistency is key, so a good mulch will provide extra protection against drought and reduce your watering requirements.

~ **Good pruning is also essential.** Get to know where the flowers will form and leave that growth. But always remove dead or diseased branches. Nurseries often provide pruning information with their fruit trees and shrubs, so do read the labels.

~ **Low yields are also caused by too many or too few nutrients.** Assess your plant's health. If the ground is good and the plant is healthy, stress it by bending branches and reduce your feeding regime. If it looks parched and sickly, give it some TLC.

~ **Net plants early** to protect them from birds and use shiny surfaces or whistling tape to scare them off.

~ **If your fruit doesn't ripen,** remove shading foliage to allow light to get to the fruits and thin the fruits by half to give the remaining ones a chance to flourish. This is often necessary on apple and plum trees. As well as giving the remaining fruit space to grow, it will prevent branches from breaking under the weight of excess fruit.

~ **Feed through the ripening process** and encourage plants to get going nice and early with a good spring mulch. If you're getting fewer fruits each year, give your plant a prune and a feed, and, if possible, move it to a more favourable position in winter.

3 top tips

- **If possible, use large containers** (at least 45cm wide) for fruit along with a soil-based (John Innes) compost – multi-purpose compost will dry out too quickly.
- **Fruit plants need** a high-potassium (sometimes called potash) feed, regularly and consistently. This is indicated by the letter 'K' in fertilisers. Look for sulphate of potash or tomato feed, liquid seaweed feed or comfrey tea.
- **Pollination involves transferring pollen grains** from the anther to the stigma. The easiest way to make this happen is by encouraging more pollinating insects, such as bees, into the garden.

'Fruit growing can be one of the most rewarding things we do as gardeners. But it helps to understand how to get the best from our fruit plants. Fruits only form once the flowers have been pollinated. Bearing this in mind makes success more likely, as people often focus on protecting the actual fruit, forgetting what's needed to encourage the plants to start flowering.'

Frances Tophill

FRUIT CALENDAR

Spring

~   Cover fruit trees with fleece if frost is forecast – remove covers by midday to let pollinating insects in

~   Water newly planted fruit regularly until established

Summer

~   Thin out plum crops to prevent branches breaking

~   Net fruit against birds but check daily to make sure birds haven't got caught in it

~   Trim strawberry runners to ensure a good crop next year

~   Tie in climbing fruit such as melons

~   Prune plums to avoid silver leaf disease

~   Pick off baby fruits from new fruit trees in the first year to help the trees establish

~   Pick fruit when ripe but not overripe as it may rot and spread disease or attract wasps

Autumn

~   Remove old leaves from strawberries and excess runners and clear away old straw at the end of the growing season each year

~   Check fruit tree ties aren't too tight after a season of growth

~   Mulch around the base of plants with well-rotted organic matter to lock nutrients and moisture into the soil and suppress weeds

Winter

~ Plant bare-root fruit trees and bushes

~ Prune bush fruit such as blackcurrants and gooseberries into an open goblet shape to let air and light into the plant to prevent fungal diseases taking hold

~ Prune apples and pears to keep them healthy and fruiting well

~ Cut autumn-fruiting raspberries down to the base to encourage strong new canes next year

~ Look out for mummified apples caused by brown rot – these often stay on the tree, so pick them off and bin or bury them to stop them reinfecting the tree in the spring.

Q&A
COMMON QUESTIONS ABOUT GROWING FRUIT

How do I stop birds eating my fruit?

There is only one way to make sure that you – and not the birds – eat your raspberries and strawberries, and that is to cover them with netting. Don't use loose netting, in which the birds can become entangled, but a proper fruit cage (tall for raspberries, lower for strawberries). It really is as simple as that, and the netting can be removed and stored after cropping so that your garden doesn't look as though it has its own tennis court.

Why doesn't my apple tree fruit?

There are several possible reasons. Firstly, if the tree isn't flowering it could be that it's a seedling, not a named variety, and so may not produce as many flowers as a fruiting variety is bred to do. Non-flowering or poor flowering could also be a result of incorrect pruning, where the flowering wood has been cut out. Maybe growing conditions are stopping flowering, so check that the tree is well watered and fed with plenty of sulphate of potash. Is the tree flowering but not fruiting? It could be that the flowers aren't being pollinated because there's no suitable pollinating partner tree nearby. If you know your variety, try to find space for another apple tree in your garden – ask a local nursery which one to choose. Lack of pollination could also be because pollinating insects have struggled to cope on the site, because it's too cold or windy. Bad weather can also impede successful fertilisation of the flowers.

Why are my autumn raspberries not cropping?

This could be a pruning problem. You need to cut all the canes on your autumn raspberries right down to the ground between late autumn and late winter. New canes grow then fruit in autumn. Some people cut off the very tip of old canes so that they fruit in summer, and take a smaller crop from any new canes in autumn. Try a high-potash spring feed.

Why were the plums on my new tree tasteless?

Newly planted trees or bushes should never be allowed to crop heavily their first year, as that weakens the plant. Give the tree a thick ring of manure or compost, then next year thin the fruits ruthlessly, leaving just a few to go on and mature. Continue to do this until the tree becomes more robust.

Why do the leaves drop off my lemon?

Although citrus can withstand low temperatures, cold can make their leaves drop. A minimum of 7–10°C will improve matters, but they dislike the hot, dry air and low light of normal living rooms. Extremes of drought or overwatering could also be to blame, and they prefer soft rainwater. If they are growing at the warm temperatures above, a winter citrus feed is also needed.

What is this woolly growth on my apple tree?

This is caused by a pest called woolly aphid, small, black-ish-brown insects covered in very obvious white or grey woolly wax. It frequently appears on apple trees in late spring and early summer, but may be present all year round. Infested trees develop swellings on twigs and branches, which can crack open, allowing infections such as canker to take hold. On small trees, remove light infestations with a stiff brush, but do this early in the season to prevent a build-up of the problem.

Why are my cherries discoloured and unripe?

It could be because the flowers have not been pollinated. Despite this, fruit may still start to develop but the flesh will not thicken as it should do, leaving a rather skinny cherry whose flesh fails to ripen and which will then start to discolour. It may be worth cutting some of these cherries in half with a sharp knife. If there is nothing solid and firm within the young stone, the fruits haven't been pollinated. Another cause could be lack of warm weather and cold night-time temperatures. In these conditions, even a pollinated cherry can become discoloured.

Why won't my blackcurrant flower or fruit?

Blackcurrant flowers can be lost to frost or cold winds. So when planting, choose a sheltered, frost-free spot or protect the flowers on cold nights by covering them with horticultural fleece. The flowers of 'Ben Sarek' and 'Ben Connan' are frost resistant.

What fruit can I grow in the shade?

It's often thought that all fruit needs full sun, but there is plenty of fruit that will thrive in even a partially shaded spot. These fruits will crop with only minimal sun: gooseberry 'Invicta'; cherry 'Morello', a sour cherry that's perfect for cooking and self-fertile; raspberry 'Glen Ample', a heavy cropping summer variety with spine-free stems; redcurrant 'Junifer', an early variety that produces big crops on one- and two-year wood, and rhubarb 'Timperley Early', which is one of the earliest rhubarbs to mature and has an Award of Garden Merit from the RHS.

Do I need to thin out tree fruits?

Apple and plum trees can produce so much fruit that the branches break. For a good crop of flavoursome fruit, it's best to thin out the fruit in early July so that they are around 8cm apart. This will also help prevent trees producing a huge crop one year, and a tiny one the next.

How do I get rid of sawfly?

It's the caterpillar-like larvae that do the damage, feeding on leaves or tunnelling into fruits. Apple sawfly live in young apples. Usually apple sawfly will only affect a proportion of the fruit, which means there will still be a harvest. Gooseberry sawfly larvae devour the foliage of gooseberries and red/white currants. Inspect leaf undersides in the centre of bushes for eggs and larvae. Sawfly may cause unattractive damage to leaves but they are not a big threat to the health of the plant. A few things that can help prevent further attacks include picking up any fallen apples to prevent the larvae tunnelling in the soil and squashing larvae by hand if you spot early infestations. Encourage natural predators such as birds.

Growing Veg

INTRODUCTION

The rewards of growing veg are obvious – the flavour, the freshness of homegrown veg and the sense of satisfaction and achievement. But there are frustrations to be faced when growing your own food – from pests and diseases to droughts and frosts. The trick is to try to reduce the chances of it going wrong. Any garden can produce vegetables, whatever its size, provided your tailor your ambitions to your circumstances. Give your veg a good start by preparing your soil well, and don't be too impatient to get sowing outdoors. Never sow seeds when the ground is too cold or too wet – these are just a few of the steps you can take to give your veg the best chance of success. The more you put into growing healthy veg, the fewer problems you will experience.

'There are two important things to remember: first, only grow crops that you enjoy eating. Obvious? Only perhaps after you've sown a row of every possible veg, in a fit of early spring passion, then find that half of them run to seed because no one in the family likes eating them. The second is to ensure that no crop goes short of water. Dryness at the roots will bring growth to a halt and may cause plants to run to seed.'

Alan Titchmarsh

FIVE COMMON VEG GROWING PROBLEMS

Slow to mature veg

A late spring and cold, damp season make veg crops slow growing. To guard against this, use early varieties where possible of peas, carrots, potatoes and so on since they mature faster. Feed your crops well, scrupulously weed and control pests to give vegetables the best chance. Use cloches and fleece at both ends of the season to keep your crops cosy.

Unripe tomatoes

It's a good idea to 'stop' tomatoes (remove the main growing tip) in August in a poor summer to encourage immature fruit-lets to fill out and ripen before the end of the season. Keep the plants a tad short of water, remove sideshoots regularly and loosen the roots with a fork, almost lifting the plants from the ground, two weeks before you plan on pulling them out, to hasten ripening of the fruit.

Carrot root fly

Carrots should be sown thinly at the outset to avoid having to thin out the seedlings later. Such thinning releases an aroma that acts like a magnet to carrot fly, which lay their eggs on the roots and results in them being eaten

by maggots. After sowing carrots, cover vegetable beds with fleece, secured at the edges, to prevent low-flying female flies reaching your crops.

Potato blight

Late blight is common in midsummer, especially after warm, wet weather. It starts with brown blotches on potato leaves, which blacken and rot until the whole plant collapses. Blight can spread to the tubers, so to save your harvest remove the top growth to soil level once the stems start to collapse. Leave your spuds in the ground to harden for a fortnight, then dig them up and eat straight away. Look for blight-resistant varieties to grow next year, such as 'Sarpo Mira' and 'Sarpo Axona'.

Bolting veg

Veg such as salad, beetroot and chard can bolt (run to seed) when their plants become stressed. This usually happens when there is too much or too little heat or the plants receive too much or too little water. The way to prevent this is to keep watering consistent. Sow successionally, so you have a constant supply of crops through the season, and try using bolt-resistant varieties such as 'Boltardy' beetroot.

'When you're just beginning, success is all-important, so stick to easy crops that give rich rewards; such as runner beans, courgettes, potatoes, salads and onions from sets. Avoid trickier crops, such as Florence fennel or cauliflowers, until you're more experienced.' **Carol Klein**

ACTION PLAN
HOW TO GROW
HEALTHY VEG

~ **Get the timing right**. From initial sowing to pricking out, planting, thinning, staking and harvesting are all improved by doing it at the right time.

~ **Prepare your soil** before planting vegetables. Add organic matter, such as homemade compost. Always use a proprietary compost, rather than homemade compost, when sowing vegetables in pots or trays.

~ **Make maximum use of space** by successional sowing – planting or sowing a new crop after harvesting the first.

~ **Allow sufficient width** between veg rows to allow a Dutch hoe to be pushed between the crops. It is still the most efficient way of keeping down annual weeds such as chickweed and groundsel.

~ **Rotate your crops** to avoid disease and maintain vigour. Try not to grow crops from the same group, such as brassicas, root crops, alliums or legumes, in the same place the next year.

~ **Use crop-protection netting** to keep butterflies off your brassicas and flies off your carrots. This will save your temper and make sure you can enjoy chemical-free crops.

~ **Grow onions from sets** – small bulbs – rather than seeds. They are easier to handle and avoid the need for sowing seeds and transplanting seedlings.

~ **Sow seed sparingly**, when sowing in pots or trays, so that when seedlings germinate they have plenty of room to develop.

~ **Water thoroughly** in dry spells, preferably with saved rainwater, to keep plants growing without interruption.

~ **Make your own liquid plant feed**, by soaking leaves of deep-rooted plants in water and allowing them to disintegrate. Comfrey and nettles are particularly good.

~ **Use any waste material** to make homemade compost. Your compost heap is the best way to provide soil health and is at the heart of self-sufficient veg growing.

PROJECT – How to companion plant

Companion planting means planting veg alongside flowers that will help ward off or confuse pests and will attract beneficial insects. It is popular with organic gardeners because it's the most natural way to get better results in the garden. And while there is no scientific explanation for why this happens, the effectiveness of companion planting will come as no surprise to gardeners who already grow a wide variety of plants. Growing a limited range of plants can lead to problems with pests and diseases, which is why vegetable gardening can be so prone to pest infestation and crop failure. The answer is to try to grow as many different types of plants as you can. Start by making a note of any plants in your garden that appear particularly popular with beneficial insects, such as hoverflies and ladybirds, as well as those plants that suffer from bad pest attacks. Plan your planting combinations carefully and you can make a real difference to the vigour and success of your plants. Here are four ways to try companion planting:

1. Repel pests – marigolds and tagetes produce chemicals that actively repel pests like aphids. Plant them in among the veg or around the edge of the patch. Marigolds are particularly effective with tomatoes.

2. Attract pests – some plants, including nasturtiums, are irresistible to aphids and other problem insects. Pests are drawn to these 'trap' or 'sacrificial' plants, which can be pulled up once they're inundated.

3. Attract beneficial insects – lacewings and hoverflies love flowering plants that produce lots of easily accessible pollen, such as zinnias, cosmos and dill, plus poached-egg plant, campanula, echinops, ivy, goldenrod, aster, rudbeckia and Californian poppies.

4. Mask the scent of plants – alliums, such as garlic and onions, release aromatic compounds that will mask the scent of other crops. Plant them with carrots to ward off carrot root fly.

SPOTLIGHT ON
TOMATO SUCCESS

Tomatoes need full sun to do well and enjoy free-draining soil that has been enriched with well-rotted compost or manure. Don't rush to get started. Tender plants will shrivel if planted out too soon. Wait until the weather is warming up before you sow, if sowing outdoors, and ensure all risk of frost is passed before you plant out your tomato seedlings – the end of May in most places – and they'll soon romp away and overtake plants set out earlier.

Common problems

~ **Blight** marks the fruits and makes them unpalatable. It's worse in damp weather, so grow in a greenhouse to keep leaves dry and avoid splashing the foliage. Outside, grow blight-resistant varieties like 'Mountain Magic'.

~ **Whitefly** sap suck and weaken your plants – grow French marigolds nearby to deter them. Alternatively, you can release the biological control *Encarsia formosa* (a parasitic wasp) in early summer.

~ **Blossom end rot** is a calcium deficiency, causing brown, corky fruit bases. It's caused by uneven watering. Never let tomatoes dry out and keep the soil damp throughout the growing season.

~ **Fruit splitting** is a result of fluctuating water levels. If you flood plants their fruits will burst, especially thin-skinned varieties like 'Sungold'. With summer downpours, it is hard to avoid this with outdoor tomatoes.

~ **Verticillium wilt** is a nasty fungal disease that causes plants to droop and die suddenly in midsummer. It lives in the soil and there's no cure – so if your soil's infected, grow resistant varieties like 'Fandango' or grow plants in containers or growing bags.

Steps to healthy tomatoes

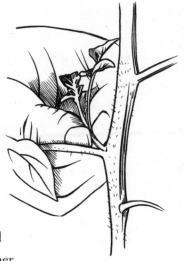

~ **Identify the type** you're
growing: check if it's a
bush variety or a single-
stem, cordon type that will
need sideshoots removing.

~ **If growing plants
outside**, choose early
fruiting types like 'Shirley',
and bush varieties such as 'Red
Alert' which produce fruit sooner.

~ **Leave space between plants** for air to circulate – don't
be tempted to plant closer than the seed packet recommends.

~ **Keep tomatoes healthy** by removing all the leaves
under fruits you've picked. It will let air circulate and so
reduce fungal problems, plus make weeding easier.

~ **Don't be greedy** – settle for three trusses of fruit and
you'll stand more chance of them ripening outdoors in an
English summer.

~ **Keep watering consistent.** Water in the morning, so
that plants do not sit damp all night, and take care to only
water the soil or compost, keeping the leaves dry. Remember, tomatoes taste better if the plants are not overwatered.

~ **Cover the soil** with a permeable membrane to help keep leaves dry. Support plants, including bush varieties, to keep leaves off the soil.

~ **Remove lower leaves**, and some higher up, to increase air circulation. Vigorous bush varieties produce too many leaves and benefit from drastic pruning to thin the foliage and encourage good ventilation.

~ **Choose the right fertiliser.** Feed only with fertilisers that are high in potassium, such as dedicated tomato feeds. Never feed tomatoes with a high-nitrogen fertiliser; it boosts leaf production, making blight more likely.

~ **Greenhouses or polytunnels** should not be allowed to become too humid. Keep them well ventilated, mop up water and, as the evenings turn cooler, use a heater to reduce condensation.

~ **Check plants regularly for blight**, from summer onwards, and dispose safely of badly diseased plants.

~ **Cut back on watering** from late summer to intensify flavours. Water just as often, but with half quantities to avoid diluting the taste.

3 top tips

- **Mulch your beds** with as much well-rotted compost as you can lay your hands on. Be especially generous where you're growing crops that will be there for a while, like brassicas and potatoes. If you don't have enough to go around everything, then skimp on the quick-growing crops like salad leaves.

- **Choose the best varieties** like star performing varieties or those that offer resistance to diseases and pests, such as carrot fly or tomato blight. Stick to reputable seed suppliers and look out for labels like 'best seller' and 'best cropper', as well as the RHS Award of Garden Merit (AGM), a guarantee of quality.

- **Use seed that is as fresh as possible**, especially for veg that is hard to germinate.

'Thin ruthlessly. It is tempting to admire seedlings filling a row, but they will be struggling to compete with each other for water and nutrients. The best weapon against slugs and snails or disease is a healthy plant, so allow room for each plant – whether a lettuce, carrot or broad bean – to grow to a strong size above ground but, more importantly, to have room for its roots to develop well. This will give you a bigger harvest over a longer period and less trouble from predators.'

Monty Don

VEG GROWING CALENDAR

Spring

~ Get your growing year started. From late winter prepare your seedbeds

~ Start sowing tomatoes and chillies under cover and early peas and broad beans outdoors

~ Start chitting early potatoes

~ Get supports ready for climbing crops

~ Keep an eye out for pests and weeds

~ Sow outdoors once the soil has warmed up

Summer

~ Start watering in earnest

~ Apply liquid feeds

~ Ventilate the greenhouse

~ Keep weeding

~ Thin out spring sowings

~ Plant out winter brassicas and leeks

~ Earth up maincrop potatoes

~ Start picking crops such as garlic, salad leaves, carrots

Autumn

~ Harvest maincrop potatoes, onions, tomatoes, chillies

~ Clear beds – pull up tomatoes, cut back Jerusalem artichokes and asparagus

~ Sow overwintering broad beans

~ Plant out shallots, onions, garlic and hardy leaves for undercover harvests

Winter

~ Harvest winter brassicas, leeks and hardy crops under glass
~ Target overwintering weeds
~ Clean pots and trays to avoid disease next year
~ Service tools
~ Check the viability of seed packets before storing

'In my experience it never works out perfectly, but growing veg is not about perfection, it is about adventure and fun and growing things you enjoy eating.'

Carol Klein

Q&A
COMMON QUESTIONS ABOUT GROWING VEG

How can I get maximum crops from my French and runner bean plants?

The best way to keep beans coming is to pick them! Harvest every few days, while the pods are still smooth, and you'll stimulate the plants to produce more. Runner beans stop cropping if they get too dry, so water during hot spells. Pinch out the shoot tips once plants reach the top of their supports, so they concentrate on producing beans instead of leafy growth. Finally, watch out for blackfly – infestations can really weaken plants, so squish by hand or spray with insecticidal soap.

How should I harvest my potatoes?

It can be difficult to know when spuds are ready, as each variety matures at a different rate. But once the foliage starts yellowing, you can lift second earlies; maincrops can wait till next month, to give them time to develop a really hefty harvest. Carefully insert a fork a little way from the plant, so you don't spear any of the tubers, then gently lever the whole thing out of the ground. Once you've collected in your bounty, double check you haven't left any tubers behind or they'll re-sprout next year.

What is the best veg for growing up walls to save on space?

Some veg thrive when trained up a vertical growing surface. Climbing beans are the obvious first choice, but also pumpkins, squash, courgettes and cucumbers will all do well if trained to grow vertically. Tomatoes, broad beans and peas will do well planted against a wall.

What should I do about tomato leaf curl?

Tomato leaf curl can be caused by temperature fluctuations (cold nights and warm days). When you grow tomatoes in a greenhouse you always find that those nearest the door are most likely to show leaf curl. Keep an eye on the greenhouse temperature with a min/max thermometer and adjust heating and ventilation if necessary. Tomatoes don't like night-time temperatures below 12°C but should recover once nights warm up in the summer. You could cover plants with fleece if temperatures drop.

What can be grown in heavy clay soil?

The first thing to do is begin adding as much organic matter as possible, such as well-rotted manure and garden compost. You don't have to dig it in – just spread it on the surface as mulch and worms will do the rest. Potatoes are good for opening out heavy soil, and all brassicas, such as cabbages and broccoli, do well. Broad beans have very deep roots that will help add organic matter, and lettuces love it. Beet crops such as chard and beetroot like clay a lot, and strawberries always flourish in it. Persevere with thick mulches of well-rotted organic material twice a year and you can grow anything you like.

Can I grow veg if I only have a small garden?

People often assume they don't have enough space for anything worth growing. But even a tiny patch, used imaginatively, can provide fresh vegetables all year. The best way to cultivate intensively is in raised beds, which can be made out of scaffolding boards. The more limited your space, the more important it is to grow the vegetables you like best. So make a list of your favourites, then work out how they could grow together, and get sowing. In fact, you don't even need a garden – just a few pots will do.

Should I sow under cover or direct?

All the standard stuff that's sown, grown and picked – carrots, radish, spring onions – is easiest sown directly. Save mollycoddling for plants that provide a crop for a whole season, such as runner beans, tomatoes and courgettes. These need to get off to a really good start, so sow them into individual modules under cover, where you can keep an eye on them until they are mature enough to plant outside.

Is there a secret to growing large celeriac?

Celeriac is not easy. Plants do best in rich heavy clay soil with a high water table. Or grow them individually in large tubs, fed and watered like prize leeks, starting the plants off early indoors with warmth. Oddly, it's also beneficial to remove some of the oldest leaves in midsummer.

Pruning

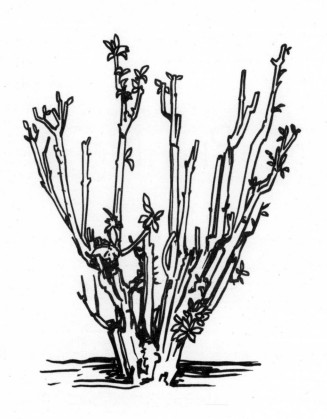

INTRODUCTION

If there's one gardening job that worries people more than any other, it's pruning. Some people worry that they don't know what to prune – or when – or how; others worry that their own plants look nothing like those shown in step-by-step instructions in books, while thrifty folk worry about paying a tenner for a plant only to be told to chop off a fiver's-worth. And everyone worries – at some time or other – that they'll cut off 'the wrong bit'. Still, when you get down to the whys and wherefores, pruning isn't really such a dark art. Judicious pruning can also turn problems into an opportunity – training fruit trees as fans or espaliers against a wall will give you twice the fruit from half the space. Pruning can also solve problems such as overgrown shrubs and slow growth.

If you leave must-prune plants to do their own thing, you're asking for trouble, so take the plunge and don't worry – the great thing about plants is that if you get it wrong, they grow back again! Most mistakes can be rectified by time, and if you know why you are pruning a particular tree or shrub before you start, then you can have in your mind a picture of what you are hoping to achieve.

Pruning comes in many forms, from simple deadheading (the removal of faded flowers to brighten up the plant and to prevent unwanted seed production), to clipping topiary specimens to create living garden sculpture, or cutting hard back to encourage more young wood. Just make sure you are comfortable with your aims before you start, and mistakes will be avoided.

'Pruning is perhaps the most daunting task facing any gardener – novice and enthusiast alike. This is probably because, aside from being seemingly complicated, it is impossible to stick back a part of a plant you wish you had not cut off!' **Alan Titchmarsh**

FIVE COMMON PRUNING PROBLEMS

Plant stems dying after pruning

Woody plants and roses may die back if the pruning cut isn't made close to a viable bud. The nearer you prune to a bud, the less chance there is of the stem dying back and the 'cleaner' the bud will break from the stem. This will also reduce the risk of disease getting into the plant.

No flowers

Hard pruning at the wrong time of year can result in a plant producing fewer flowers, or none at all, if you cut off all the shoots that should have carried the next crop of blooms. This is a common problem with hydrangeas and other large shrubs that are repeatedly pruned back hard, as they never get the chance to produce flowering shoots. Make sure you prune at the right time of year for that particular plant, bearing in mind when it naturally flowers. The general rule is to prune shrubs and climbers that flower before June immediately after flowering, and those that flower after June in winter.

Poor fruiting

A plant that hasn't been pruned correctly can end up as a tangle of branches that doesn't produce flowers or fruit. Fruit trees and bushes look and crop better if light can get into the centre of the plant and air can circulate. It also helps to prevent pests and diseases. Remove any shoots that cross over and thin out any of the rest that are too close together. After this, prune regularly to maintain the plant's health and vigour.

Orange spots on branch ends

These are a sign of coral spot, a fungus that usually colonises dead stems but can attack live twigs and branches, causing them to die back. The spots are pale orange in dry periods or bright orange when damp. The fungal spores are dispersed by water, either rain or watering, and enter trees or shrubs through pruning cuts or damage to branches. Acers, elaeagnus and magnolias are especially susceptible. In dry weather, cut out the infected stems, right down to the collar or swelling at their base, where they arise from healthy stems and branches. Bin all infected material to avoid spreading the fungus.

Lots of weak stems

If you don't prune out weak stems regularly, you'll end up with a plant that has lots of thin, spindly branches and uneven growth. Even up the growth by pruning back the weak stems, cutting them shorter than the strong ones. Where weak stems have grown from the base of the shrub, cut them out completely to promote strong new growth.

'Pruning is made easier and safer when the correct tools are used. A common problem is to use a tool too small for the size of the branch, possibly damaging the tool, the branch and the user. A sharp cutting edge is also very important, otherwise the stems can become "chewed".' **Matthew Wilson**

ACTION PLAN
HOW TO PRUNE CORRECTLY

~ **Always cut back** to just above a bud, choosing one that points in the direction you want the new shoot to grow. For example, prune free-standing shrubs to outward-facing buds to form a well-shaped plant with an open centre. When pruning wall-trained fruiting or flowering shrubs, choose a bud that points out across the wall to fill a gap, instead of outwards over the path where it'll just be in the way. If you can't see an actual bud, prune to a leaf joint, as that's where the plant can produce a shoot bud, even if you can't yet see one.

~ **Clean secateurs before pruning.** If you have been pruning out infected tissue, there is a danger that you could spread disease to healthy plants unless your secateurs have been properly cleaned and disinfected.

~ **Cut out any damaged branches**, so that disease doesn't set in and affect the health of your plant. Cut back to an intact stem, just above a healthy shoot or bud that can support strong new growth. Where this will result in an unsightly or oddly shaped branch, consider cutting back to a point lower down, to promote regrowth from further down the plant.

~ **Prevent rubbing stems**. Closely spaced or crossing stems and branches tend to rub against each other when they move in the breeze. This can wear away their protective bark, creating an easy access point for disease. So cut out any shoots showing signs of damage. And when doing any pruning, cut nearby branches back to buds facing in different directions, so the resulting shoots will grow in different directions.

~ **Remove stems that compete**. Left to their own devices, established trees, shrubs, fruit bushes and climbers may produce several stems, each of which is weak as a result of competition for space and light. Prune out one or more of these competing stems, so those that remain grow more robust and are better able to support their own weight, especially when laden with flowers and fruit.

~ **Don't leave snags.** If a branch does not include visible or dormant (often invisible) buds it will not be able to resprout. This results in a 'snag', which will die back to the bud below. Sometimes you can't tell where the dormant buds are, so you need to guess. Check your cuts in spring after leaf growth and cut off any snags then, cutting back to the nearest healthy shoot below. Snags are prone to disease.

~ **Tackle the shrub's centre,** don't simply clip away at the edge of your shrubs to keep them within bounds. This will lead to a dense, twiggy and unproductive centre with poor air circulation, which encourages pests and diseases to thrive. Instead, be brave and start by removing any stems and branches you need to in the centre of the shrub, then thin out the remaining outer stems to let in as much light and air as possible.

~ **Cut out signs of disease**. Most well-pruned and healthy shrubs, climbers and fruit bushes can shrug off diseases. However, it's worth inspecting them carefully for signs of fungal infections, such as coral spot, leaf spot, rust and mildew, which can overwinter on bare stems or old snags, then infect new shoots next season. Prune out all infected stems, cutting back to healthy wood.

PROJECT – How to tackle an overgrown shrub

Giving overgrown deciduous shrubs like forsythia or viburnum a light trim encourages them to form a dense mass of woody stems that carry few flowers. Overgrown shrubs need more hard pruning in order to perform well.

1. Cut out a third of the thickest, oldest stems from the base in winter, using a pruning saw or loppers.
2. Cut back any crossing or damaged branches.
3. After flowering, reduce the remaining wood by half.
4. Mulch the plant well after pruning. Prune annually after this to keep the plant under control.

SPOTLIGHT ON
COMMON PRUNING
MISTAKES

Pruning too much

If a plant is growing well, is not full of dead or diseased wood and has a pleasing shape, leave it alone. Don't butcher your plants if they are too big for their space. Remove them completely or move them to a spot where they can grow freely and replace them with something of a more modest stature that better suits the space. Always avoid cutting back too many of the branches that are carrying flower buds, usually the two- or three-year-old stems. This is especially important in spring.

On the other hand, summer-flowering shrubs, climbers and bush roses bloom on the tips of growth made during the same growing season, so can be cut hard back in the dormant season without any detrimental effect on flowering.

Not pruning enough

Spring-flowering shrubs such as philadelphus and weigela will become congested and overcrowded with non-flowering stems if left to their own devices. Removing a few of the older branches after flowering will keep them youthful. Wisteria will run rampant if not pruned in summer to reduce the spread of those long, snaking growths. Cut them back to 30cm in July or August and in January shorten all sideshoots to finger length to encourage the production of flowers.

Not pruning with the right technique

Prune above a node (leaf joint) and the bud immediately below the cut – in the leaf 'axil' – will grow away well. Cut too far above the leaf joint and the stem will die back to the bud – and probably beyond, risking infection. Make your pruning cuts at an angle of 45 degrees, sloping away from the bud, to help shed water and reduce the likelihood of disease gaining a hold.

Not removing die-back

If a stem is dying back, the infection can continue to advance down the stem unless that stem is cut out well back into healthy tissue further down the plant.

Pruning at the wrong time of year

Trained fruit trees will produce more leafy growth if they are pruned in winter. Shorten the sideshoots in summer and you will encourage the production of fruiting spurs, which will result in a better harvest. Plum trees, in addition, are susceptible to the debilitating fungal disease silver leaf, which is most likely to attack surfaces that are pruned in winter. So always do your fruit tree pruning, if it is necessary to keep the tree shapely, in summer.

3 top tips

- **Always use pruning tools** that are designed for the job you are doing. This will help you to make clean, healthy cuts. Cut branches that are bigger than 1.5–2cm with loppers or a pruning saw and use long-handled tools to increase leverage and reach.
- **Avoid pruning** when plants are under stress, especially during drought. If dry weather persists and you need to get on with the job, water the plants really well before you prune them.
- **Make sure that plants are watered** and fed well immediately after pruning to give them everything they need to produce new shoots. Make a habit of mulching all trees, climbers and shrubs after you have pruned to set them up for their burst of extra new growth in spring.

PRUNING CALENDAR

Spring

~ Most evergreens

~ Winter-flowering shrubs such as *Viburnum* x *bodnantense*

~ Tender shrubs such as lavender

~ Summer-flowering shrubs such as fuchsia and buddleia

~ Spring-flowering shrubs – after flowering, in late spring or early summer

Summer

~ Early summer-flowering shrubs – after flowering, such as rhododendrons, azaleas and philadelphus

~ Trained fruit and hedges

~ Spring-flowering climbers

Autumn

~ Summer-flowering jasmine

~ Summer-fruiting raspberries

~ Plants that bleed sap in spring and summer, such as birches

~ Perennials

Winter

~ Overgrown shrubs

~ Apples and pears

~ In late winter – most roses, clematis and ornamental grasses

~ Autumn-fruiting raspberries

~ Fruit bushes

Q&A
COMMON QUESTIONS
ABOUT PRUNING

How can I reduce the size of a large viburnum?

Viburnum tinus is a very robust and vigorous shrub that can cope with being pruned to ground level and will regrow readily. Prune it in late spring, once there is no danger of frost and make sure to check for birds' nests before you begin pruning. Work out what height you would like it to be in the long term, then cut back to at least 30cm below that, to allow for strong regrowth. Pruning in spring also means you won't have to look at a bare shrub and pruning cuts for too long. After pruning, feed with a general slow-release fertiliser, mulch with garden compost and water during dry periods. In future years, cut it back every spring to keep it within its allotted space.

When is the ideal time to prune a plum tree?

Plum trees should be pruned between June and August. This is because they're susceptible to silver leaf, a fungal disease that causes leaves to turn silvery and branches to die. The spores are more widespread in autumn and winter, and can enter through pruning cuts, so prune only in summer to avoid infection. Also sterilise all pruning tools between cuts with a garden disinfectant.

How do I prune a tall buddleia?

Buddleja davidii benefits from being hard pruned in March on a regular basis. Take all stems back down to 30–60cm from the base. This may appear to be brutal, but it will encourage the plant to put on growth lower down, creating a good shape and plenty of flowering shoots.

How should I prune an overgrown montana clematis?

Usually montanas should be shaped after flowering and that's enough maintenance to keep them looking good, but if you really want to clear out the plant, then you can either prune out the congested parts or renovate the whole thing. This can feel a bit brutal, because after flowering you'll need to cut it right down to the old wood.

Give it some love in the form of organic mulch and go back to light pruning for the next few years. Before you get your pruning saw out, make sure that you check plants carefully for birds' nests.

Why is my mophead hydrangea not flowering?

This is often down to pruning or growing conditions. Mophead hydrangeas need dappled shade or sun, in moist but not water-logged soil, and protection from scorching sun and cold wind. Leave the old flower heads in place over winter, then cut back to the strongest pair of buds just below, after all danger of frost has passed. At the same time, cut out two or three of the oldest stems at the base to encourage new flowering wood to grow. Don't cut the whole plant to ground level or it won't flower until the following year.

Avoid nitrogen-rich fertilisers, which encourage lush growth at the expense of flowers. Instead, apply a high potassium feed in spring to encourage blooms. Mulch in spring too, as this helps to retain moisture in the soil, which is needed for flowering. Use a 7cm layer of well-rotted organic matter, such as homemade compost.

Does wisteria need pruning twice a year?

Yes, if you want the best flowering display. Around two months after flowering, in July or August, cut back sideshoots to within five or six buds of the main branch to control leaf growth and encourage short flowering spurs. Then in winter, cut the stems back harder, to two or three buds from their base. Plants will flower without pruning as they do in the wild, but spur pruning will give you a more abundant display.

What is coppicing and why is it done?

Coppicing involves cutting back a tree or shrub to the base, so that it sends up vigorous new shoots. It has long been used on trees such as hazel to produce a crop of straight poles. It can also cause plants such as paulownia to produce huge, lush foliage.

Greenhouses

INTRODUCTION

You don't have to be a gardener for long before you yearn for a greenhouse. It's the one thing that will allow you to enjoy your pursuit when the weather outdoors is inclement. Plus, it opens up opportunities, such as sowing the seeds of tender flowers and crops that would otherwise have to be bought as plants later in the season, and offers a protected environment that allows you to grow tomatoes, peppers, cucumbers and melons regardless of capricious weather.

The initial outlay will be mitigated by your ability to save money – plants raised from seeds and cuttings are cheaper than nursery-grown ones, and you'll be able to grow a wider range than usual. Decorative pot plants can turn your little palace of glass into a conservatory. Even if all you have is a glazed porch or lean-to, you'll wonder how you ever managed without it.

Getting plants to grow successfully in your greenhouse is all about balancing heat, light and airflow. If you are new to greenhouses then things like watering, heating and shading can seem a bit daunting, but actually they are easy. You will quickly begin to get a good idea of when things need to be done by keeping an eye on the weather and watching what's growing.

'The more ventilators a greenhouse has, the happier the plants – and the gardener.' **Alan Titchmarsh**

FIVE COMMON GREENHOUSE PROBLEMS

Too hot

When the weather warms up, ventilation is key to making sure that plants stay happy and healthy. Some plants such as chillies, tomatoes or melons can take more heat; however, too much can cause damage. Make sure you have a thermometer in your greenhouse to monitor temperatures – damage usually occurs when the heat rises to over 27°C. Roof windows are an ideal way to let hot air escape. Another option is side vents – these are a good way to make sure there is air movement around plants lower down. And don't forget your greenhouse door can also be a temperature regulator. Make sure you have a hook or weight so it can be held open. Some netting across the door will stop nosy wildlife trying to enter overnight.

Shading, too, is vital between May and September to prevent plants from being scorched by the magnified effect of the sun. The door can be opened, but high-level (ridge) and lower (eaves) ventilators are vital to ensure that temperatures can be kept at an acceptable level and that humidity can be controlled by a healthy flow of air.

Too cold

Tender plants will not thrive beneath certain temperatures. Keep an eye on day and night-time temperatures so you can manage the heating. Electricity is by far the most efficient way of heating a greenhouse, since it offers dry, controllable heat.

There are lots of greenhouse heaters that are fitted with thermostats to ensure efficiency and economy and they will allow you to control the heat to your desired temperature.

Unless you want to grow tropical plants in winter, you won't need to heat your greenhouse to a high temperature. Most plants, such as citrus, pelargoniums and tender fuchsias, will be happy if the temperature doesn't drop below 7°C, though if you can keep them warmer than this they will continue growing. Freedom from frost is critical – unheated greenhouses are useful in spring for the shelter they provide but in the dead of winter they offer little useful protection.

Fungal diseases

When it comes to diseases such as mildew and botrytis, adequate air circulation will keep outbreaks to a minimum. Removing dead leaves and flowers from plants will help and is a practice that every greenhouse gardener should adopt. Dampness and humidity are ideal conditions for fungi and moulds, so water sparingly in winter. Add a mulch of grit to your pots, to keep the crown of your plants as dry as possible, and open the vents on warmer days to get the air moving.

Red spider mite

This sap-feeding mite is a common greenhouse pest. You may notice leaf drop or mottled leaves. It thrives in warm, dry conditions so is most prevalent between

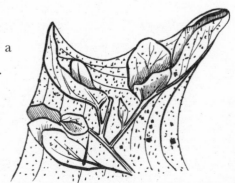

March and October. Look out for the telltale webbing on leaves and tiny yellow mites as well as white eggshells on the underside of leaves. Check plants regularly so that you can take early action. Spraying plants with water to raise humidity can help prevent attacks. Don't overcrowd plants. Try a biological control specific to red spider mite. Remove any heavily infested plants from the greenhouse and deep clean the greenhouse over winter to get rid of any overwintering mites.

Whitefly

It'll be obvious if there's a whitefly infestation in your greenhouse. You'll spot clouds of white flies around your plants and on the backs of leaves. These sap feeders leave a sticky substance called honeydew on leaves, which encourages a black sooty mould to grow. This can weaken plants. Organic solutions include catching adults by hanging up sticky traps, growing strong-smelling plants such as marigolds alongside crops like tomatoes as a deterrent and spraying the underside of leaves with a soap-based spray that contains fatty acids to kill any emerging adults. You could also try a biological control.

'I think it's a really good way to learn how things grow throughout a season if you have to go and check on them every day to see whether they need a drink. Picking the plant up and checking the weight is a good indicator and of course you will water more from late spring to autumn and less over the winter months.'

Adam Frost

ACTION PLAN
HOW TO KEEP GREENHOUSE PLANTS HEALTHY

~ **Keep on top of watering.** When it comes to watering, a tap installed inside the greenhouse will save you from lugging a hosepipe or watering can up and down the garden. Automatic watering systems are now available at a reasonable cost and can be rigged up to a small battery-operated computer fitted to the tap. When you are absent for long periods of time they can be a useful alternative to trying to find a helpful neighbour who knows how and when to water plants. A good tip is to keep a tub of water in the greenhouse, which is useful for dunking plants into and helps to keep the humidity up. Plants will need more watering from late spring to autumn and less over winter.

~ **Use capillary matting.** This is not vital but can be useful under seedlings and younger plants as it is important to make sure that they don't dry out.

~ **Damp down the greenhouse.** Create humidity by watering paths and staging during hot spells.

~ **Protect plants from frost.** Keep fleece handy to wrap around anything tender if the weather gets very cold. If you've got internal blinds, putting them down at night helps hold in some warmth and they can be rolled back again in the morning. Some people like to insulate their greenhouses with bubble polythene. This is a relatively cheap and easy method, but bear in mind that it does block out a certain amount of light.

~ **Choose practical benches.** Maximise space by choosing the right benches and shelving so you can really fill your greenhouse with plants and seedlings. Metal is easy to clean, sturdy and less likely to harbour pests. Slatted benches and shelves are best because they allow good air circulation, excess water to drain away and maximum light.

~ **Get your shading right.** Shading is another useful way to help keep your greenhouse cooler during hot spells and it will help to keep direct sunlight off foliage too. Still, it's important to find the balance between having enough shade, yet not blocking so much light that it slows down plant growth or delays crops ripening. There are quite a few different options to create shade, from blinds fitted on the inside or outside, to the cheaper option of mesh or netting that can be clipped on when required. Shade paint is another option; this can be applied in layers as the weather gets hotter, then washed or brushed off in autumn. Make sure you check the instructions on the tin to see if the paint is suitable for your greenhouse.

PROJECT – How to use biological controls

Biological controls are very effective in controlling some of the worst garden pests, such as slugs, vine weevil, ants and leatherjackets. They are perfectly safe to handle and have no detrimental impact on other species or on soil chemistry. Nematodes are microscopic worm-like creatures that occur naturally in the soil and rely on soft-bodied prey to survive. They are specific to the pest, so you can't use a nematode for slug control on vine weevil, for example. The formulations are supplied chilled, in powder form, and watered into the soil or pot where a pest issue has built up. They work best soaked into well-drained soil or compost where they can swim to their hosts. Mail-order suppliers are best as they are dispatched to you when soil temperatures are right for success. Some formulations that last a bit longer are now available in garden centres.

1. Look for signs of pest damage and identify the culprit. For example, if leaves have notches eaten out along the edges, this indicates that vine weevils have caused the damage.
2. Take the packet of nematodes out of the fridge and read the instructions. While using nematodes is easy, it is important to follow the detailed instructions to ensure success.
3. Add the nematodes to a watering can filled with water. It is important to suspend the mixture in the water, so stir thoroughly with a long stick or spatula and use straight away.
4. Apply the mixture using a coarse rose on the watering can. You are targeting the grubs lurking in the compost of a container or soil in the garden, so give it a good soak.

SPOTLIGHT ON
CLEANING YOUR GREENHOUSE

Keep your greenhouse gleaming to ensure plants and crops stay healthy. The best time to clean your greenhouse is when you have the fewest plants growing in there, which for most people is usually winter.

~ **Move any remaining plants to a sheltered spot** while you clean, so you don't have to work around them.

~ **Brush down and sweep** the inside of the greenhouse, getting into every nook and cranny. Aim to get rid of all the accumulated dirt and debris, as this can harbour pests and diseases.

~ **Clean benches, shelving and staging** thoroughly with disinfectant or detergent.

~ **Wash the glazing,** both inside and out, to improve light levels. Use a hose, or a bowl of warm soapy water and a sponge. Clear any algae and moss from the roof in particular.

~ **Clear the guttering** – run your hand along the inside to remove any debris that has fallen in. A wire coat hanger makes a useful tool to unblock the top of downpipes. Rinse through when you've finished and check everything is flowing freely.

~ **Drain the water tank,** if you have one, then scrub the insides to help stop algae build-up and waterborne diseases. Do the same with any water butts on the greenhouse downpipes too.

---

3 top tips

- **Stop diseases spreading** by putting infected material in garden waste bags – don't compost it. Although diseases can't be cured, some can be treated with fungicide.
- **Select disease-resistant varieties** wherever possible and buy healthy plants from a reputable source.
- **Adopt good gardening techniques,** such as watering onto soil rather than foliage and spacing plants out to improve air circulation, to minimise disease spread.

GREENHOUSE CALENDAR

To reduce problems in your greenhouse, it's important to balance heat, light and airflow as well as keep everything clean to avoid pest and disease outbreaks. Here are some seasonal tasks to help you keep on track.

Spring

~ Protect greenhouse sowings of beans, peas, mangetouts and sweet peas from hungry mice

~ Open greenhouse vents on sunny days to stop humidity building up but close them at night when temperatures drop

~ Remove dead flowers and leaves to prevent fungal diseases

~ Water crops regularly as the weather warms up

~ Check plants for pests, especially on shoot tips and the underside of leaves

~ Prick out seedlings as soon as they get their first true leaves to prevent overcrowding and fungal diseases

~ Use biological controls if needed

Summer

~ Increase shading and ventilation to keep temperatures down on hot days

~ Water greenhouse tomatoes regularly to prevent splitting and blossom end rot

~ Damp down the greenhouse floor each morning to increase humidity

~ Sweep greenhouse staging and floors to reduce debris that can harbour diseases

Autumn

~ Tidy up the greenhouse, getting rid of any old compost or debris that could hide unwanted visitors

~ Take down shading as light levels fall

~ Look out for frost forecasts and prepare to bring in your more tender plants

~ Wash greenhouse windows to let in more light

~ Line greenhouse glazing with bubble insulation as soon as temperatures start to drop

~ Inspect any plants you are planning to move into the greenhouse for the winter to make sure they're free of pests

Winter

~ Pick dead flowers regularly from plants overwintering in the greenhouse, such as pelargoniums

~ Monitor greenhouse temperatures with a min/max thermometer to make sure heaters are working properly

~ Water plants sparingly to keep the greenhouse dry, which should reduce outbreaks of disease

~ Check overwintering plants for red spider mite and other pests and take action if necessary

~ Remove any dead leaves from overwintering plants to prevent fungal diseases

Q&A
COMMON QUESTIONS ABOUT GREENHOUSES

What can I grow in my shady greenhouse?

Don't despair, because a lightly shaded greenhouse tends not to overheat in summer. You'll need to experiment to see what works. If cucumbers grow well, melons might ripen. Try achocha and cucamelon, although their growth is rampant. You could extend a short growing season by starting broad and French beans, plus carrots, lettuce and spinach early. In July, sow rocket, hardy lettuce, mustards, mizuna, pak choi and other leafy greens for winter. You could also try growing ferns or alstroemerias.

Where should I put my thermometer in the greenhouse?

Temperatures vary in different parts of the greenhouse. Ideally, for an average overall temperature, put a min/max thermom-

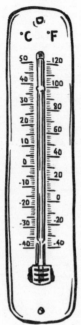

eter in the middle of the greenhouse, out of direct sun, away from cold draughts, and among the plants. If you're propagating, then lay the thermometer on the bench to get a reading that would be roughly the same as in your seed trays.

Monitoring temperatures is particularly import-ant in the heat of summer and cold of midwinter. Temperatures shouldn't be above 27°C in summer, as this damages temperate plants – open the vents early, damp down and use shading to keep things cooler. Ideal winter temperatures depend on the type of plants you're growing. Some are very tolerant and need little heating, while more tender plants may need to be kept above 5°C.

I don't have much space. What could I grow in a mini greenhouse?

Mini greenhouses allow a lot of flexibility and can be used right throughout the year. The shelving means you can't grow anything too tall, but it can be used to raise young plants, then once the tender plants have been moved outside, use that space for crops in containers that benefit from being sheltered. These include cut-and-come-again lettuces and salad leaves, dwarf chillies, stump-rooted or small-rooted carrots and beetroot, salad onions and shorter varieties of tomatoes. Growing salads this way means the leaves you harvest are very clean, as very little soil splashing takes place. Not only does the polythene provide additional protection for crops, it also means conditions can get very warm inside, so regular watering is crucial – even daily in summer.

Greenhouse or potting shed? Which is best all year?

If you only have room for one, I'd definitely choose the greenhouse. When fitted with staging a greenhouse can be a potting shed, but not vice versa. Even unheated, it can be used in winter to shelter slightly tender plants such as cannas, agapanthus, olives, ginger lilies and potted dahlia tubers. Just drape fleece over plants in long freezing spells. Cuttings of hardy plants often root better under unheated glass, where you can regulate their watering. Just keep plants well spaced and open vents a crack to provide air circulation.

In the new year, you can make early sowings of crops like broad beans, chard, lettuces and parsley in a greenhouse. Then, as the weather warms, you can sow all kinds of veg and flowers for the garden. By mid to late spring you could be growing tomatoes, basil, cucumbers and physalis under glass for tasty summer crops.

How do I tackle pests?

A greenhouse has its own range of pests – red spider mite, mealy bug, whitefly, vine weevil and scale insects. These can be controlled with bought predators and, in the confines of a greenhouse, this biological control is usually more effective than it is outdoors. Keeping your greenhouse clean will also help prevent pests and diseases.

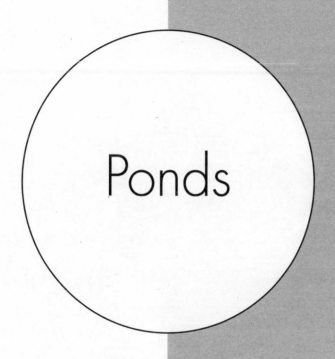

Ponds

INTRODUCTION

Ponds bring a huge range of benefits to the garden such as wildlife, colour from the plants and the calming sound or look of water, but the potential problems from green water or slimy weeds can be offputting. For a successful pond, it's important to consider where you site it. A warm sunny place is ideal for most wildlife and for your pond, too, encouraging good plant growth and warm water. But shade over some of the pond will help to reduce algae.

Avoid siting it under mature trees, not just because of the shade, but because in the autumn the water can become choked with falling leaves. As they decompose, these will absorb the oxygen in the water, killing pond life and dramatically decreasing the amount of species the pond can support. Siting it near vegetation, such as low shrubs, will provide shelter and cover for birds and small mammals. Including your pond within a border, or the veg plot, will also help you in the fight against garden pests and problems as the beneficial wildlife will be right where you need it among your plants or crops.

Unlike any other garden feature, a body of water is beautiful and multi-functional. Ponds quickly evolve complex ecosystems that support everything from micro-organisms and pond critters to the birds and amphibians that feed on them. From the tiniest pond in a barrel, to a sprawling lake, there is a style to suit every garden, and they'll all increase diversity.

'Contrary to popular belief, a pond is not a work-free zone. You still have to do a bit of planting, lifting and dividing, and even weeding. A pond is like a perennial border where everything is done under water. Instead of looking after the soil by mulching and feeding, you have to take care of the water quality.' **Alan Titchmarsh**

FIVE COMMON POND PROBLEMS

Green pond

Warm weather and/or excess nutrients encourage algae to proliferate. Newly filled ponds often turn green due to chemicals in tap water, while other causes include fertiliser or organic matter draining into the pond, and fish faeces in the water.

The water should clear once that mystical stage known as 'a balance' has been achieved. This balance is the result of sufficient submerged oxygenating plants being given a chance to work their magic, and some of the pool's surface being covered by water lily leaves or other floating aquatics, providing shade. Grow aquatic plants to shade at least a third of the pond's surface. Add plenty of submerged oxygenating plants and watercress to take up nutrients and 'starve out' algae.

Blanket weed

When water in ponds warms up, blanket weed – a form of green, filamentous algae – will start to romp away. Its dense growth, made of many fine strands, can quickly choke a pond compromising the movement of fish and other pond life. You can use a net to fish it out, but the simplest method is to use a bamboo cane, pushing the end of it into the blanket weed and then twisting to twirl the strands together before pulling them out. Work from the edge of the weed, taking out small amounts each time, rather than poking the cane into the middle of a dense clump, as this can trap fish and wildlife. You can also deter blanket weed with barley straw – buy as rolls or pads and float them in the pond.

Cloudy water

Pond water is cloudy when it contains large quantities of stirred-up silt and microscopic algae. The key to minimising this is creating a balance between the plant and animal life, so you rarely need to clean out your pond. The number of oxygenators, marginals, deep marginals, floating plants and water lilies should relate to the surface area of the pond. Aim to cover at least half the pond with foliage. If you have fish in your pond, take care not to overfeed them, as this will add excess nutrients to the water. Sometimes, adding water snails can help.

Mosquitoes

The best solution is fish, which will give you season-long relief. Goldfish are cheap and cheerful, but they will also eat tadpoles and almost any other wildlife. Sticklebacks and other native fish are less voracious, but difficult to find in aquatic retailers, and shouldn't be collected in the wild.

Dragonfly larvae are prolific mosquito hunters, although they won't reliably control large numbers. It's still worth trying to attract them to your pond, so grow a good range of aquatic plants, including marginals, submerged and emergent plants, and leave an area of long grass close by your pond for the newly-emerged adults and other insects that your dragonflies can feed on.

Mosquito larvae hang from the water surface, breathing through a snorkel-like tube, so disturbing the water can prevent them from breathing. Installing a fountain or waterfall will do this, as well as aerating the water, which is beneficial to both fish and wildlife.

Finally, if you've got mosquito larvae in your water butt, you can prevent them from breathing by floating a thin layer of regular cooking oil on the surface.

Stagnant water

Site your pond in a sunny spot. Ponds need a bit of shade to reduce algae problems but the entire water ecosystem needs sunlight to 'work'; without it, water plants (such as water lilies) fail to flower and oxygenators cannot make oxygen, so pondlife including beneficial micro-organisms die and the water turns stagnant. If your pond is larger than a container pond, using a filter will help with stagnant water.

'Both for us and for the creatures that will make it and its environs their home, a successful pond needs plants at every level: under the water, on the water's surface, in the water around its edges and in the damp and boggy areas that hopefully will surround it.' **Carol Klein**

ACTION PLAN
HOW TO KEEP YOUR POND HEALTHY

~ **Top up water** as it evaporates in summer, causing the level to drop. Ideally use rainwater from a butt around once a week or as needed. If you only have tap water available add smaller amounts more often or fill it in a container first and leave it for 24 hours before adding it to the pond.

~ **Clear decaying vegetation,** which increases nitrogen levels, turning water green while building up a layer of sediment at the bottom, which can make the pond smell. Remove dead or dying foliage frequently.

~ **Remove** pond algae, such as blanket weed, which spreads quickly and can choke a pond.

~ **Assess marginal plants** in summer and early autumn. During the summer, pond plants can become overgrown and create too much shade on the pond. Those that have outgrown their allotted space can be reduced in size simply by splitting the rootball with a bread knife and replanting the divisions in aquatic baskets. Then, move on to the oxygenators, removing excess growth and aiming to leave 50 per cent of the water surface covered. Deadhead flowering plants as you would in the garden and snip off any brown or dead leaves. Take out some of your submerged oxygenating plants if these have become overgrown – you need around four bunches for each square metre of pond. Leave the plant material you remove around the edges of the pond for at least 24 hours, so creatures caught up in the foliage and roots can escape back into the pond.

~ **Promote species diversity** – to support a broad variety of wildlife, include as wide a range of pond plants as possible. Leafy marginals offer shelter for many species, while reed-like plants provide a home for dragonflies, and water lilies an egg-laying site for water snails.

PROJECT – How to repot aquatic plants

Pond plants will quickly root through the mesh base of an aquatic container. Repotting extends the life of the plant and allows the clump to get bigger. Spring is the best time to do this as the plant goes straight into growth afterwards. Use a specialist mesh basket to allow water to soak in and surround the roots. The compost in the container must be soil-based and low in nutrients so as to deter algae, which quickly multiplies in water with high nutrient levels. After potting, soak the whole container thoroughly and gently lower it into the margin of your pond until the rim is level with the water surface.

1. Remove the old pot. It may need to be cut away from the roots with a knife or scissors. Carefully remove any dead leaves and weeds from the plant.
2. Gently tease out the roots ready for potting. Then add a layer of low-nutrient, aquatic compost into the base of the new mesh pot.
3. Place the cleaned plant into the centre of the pot and firm it into the compost. Pot up, leaving the compost level at least 3cm lower around the edge of the pot.
4. Pour washed gravel around the plant then dress the surface of the mesh pot with a layer of larger stones to weight down the compost once it goes into the water.

SPOTLIGHT ON
PLANTS FOR A
SUCCESSFUL POND

~ **Make sure you have a good mix** of three types of plant: oxygenating, floating and marginal, to provide greenery for all depths of water. Oxygenators grow mainly underwater, producing oxygen and absorbing impurities. Floating and emergent plants cover the surface of the water, giving shade below. Marginals, as their name implies, are happiest at the edges of the pond, with their toes in water or damp soil.

~ **Take into account the eventual size** and spread of each plant when you decide which to buy. Pond plants are fast growing, so keep numbers to a minimum and avoid rampant varieties. Sticking to the same rule of thumb you may apply to the rest of the garden, select a few favourites for groupings of three rather than lots of single specimens.

~ **When buying native plants**, make sure they are from a reliable, sustainable source. There are specialist nurseries that, by propagating, can ensure no 'alien' imported or wild plants are sold on.

~ **Water plants generally prefer full sun** but most of them will grow well in light shade. In fact, a position with less sun will actually curb their growth, while regular thinning should prevent even the most vigorous of water plant species taking over your pond completely.

~ **Create a natural-looking edge** to the pond with marginals. Examples include: *Myosotis scorpioides* (water forget-me- not), *Iris sibirica* (Siberian iris) and *Caltha palustris* (marsh marigold).

~ **Include surface floating plants** that are not rooted into the base or side of the pond. Examples include: *Hydrocharis morsus-ranae* (frogbit), *Callitriche palustris* (starwort) and *Stratiotes aloides* (water soldier).

~ **Include submerged oxygenators,** such as *Potamogeton crispus* (curled pondweed), *Ceratophyllum dermersum* (hornwort), and *Myriophyllum spicatum* (spiked milfoil), to grow below the water.

~ **Use bottom-rooted plants** that grow up to the water surface to get the maximum light. Examples include: *Nymphaea* (water lilies), *Aponogeton distachyos* (water hawthorn) and *Nymphoides peltata* (fringe lily).

3 top tips

- **Provide access for wildlife.** Use cobbles or rocks, carefully stacked on marginal shelves, so that they just break the water surface and create a bridge where frogs and other creatures can enter the pool, and birds, insects and small mammals can come for a drink.
- **If you can't fill your pond with rainwater,** fill containers with tap water and leave for 24 hours. Tap water is generally alkaline – this method will make the water slightly acidic and more comparable to pond water.
- **Add barley straw** to ponds. Barley straw limits algal growth by absorbing excess nitrogen. Remove straw after six months or when it has turned black.

POND CARE CALENDAR

Spring

~ Buy pond plants if needed before the growing season as the weather begins to warm up

~ Divide and repot plants if needed

Summer

~ Top up water as it evaporates in summer, causing the level to drop. Ideally, use rainwater

~ Clear decaying vegetation, which increases nitrogen levels, turning water green while building up a layer of sediment at the bottom, which can make the pond smell

~ Remove pond algae, such as blanket weed

~ Reduce the volume of oxygenators, such as hornwort, if they fill more than a third of your pond

~ Deadhead flowers on water lilies

~ Feed water lilies

~ Cut back plants if necessary

Autumn

~ Place netting over ponds before the leaves start falling, to catch them and prevent them from damaging the pond ecosystem. Simply weigh down plastic mesh with bricks and clear it every few days

~ Cut back and compost dead pond plants before they rot down

Winter

~ Keep ice and leaves off ponds by scooping out dead foliage and floating a ball or piece of polystyrene on the surface of the water to stop it freezing over

Q&A
COMMON QUESTIONS ABOUT PONDS

Which plants can I use to conceal a pond liner?

Around the outside of the pond, grow ground cover to soften the edges. If the area is damp, consider *Gunnera hamiltonii*, a diminutive evergreen with heart-shaped, leathery leaves, or *G. magellanica*, which is evergreen in mild winters. If it's dry,

Campanula portenschlagiana, with its mauve, bell-like flowers, will flourish. Also try helianthemum, the evergreen sun rose, or bergenias, such as big, bold 'Bressingham Ruby'. In the water, try marginals, such as *Alisma plantago-aquatica* with fresh green leaves and small white flowers, and *Anemopsis californica* with cream, honey-scented flowers in summer. *Caltha palustris* will grow on shallow shelves, as will elegant, pink *Hesperantha coccinea*.

How do I attract wildlife to my pond?

The success of a wildlife pond is all about achieving a healthy ecosystem, and plants are the biggest contributors to this. Ideally, create habitats from plants emerging from the water near the pond's edge, through marginals and bog plants to drier ground with brambles, flowering shrubs and trees some distance away (allowing enough distance to prevent leaves falling into the pond).

Most important are plants that stick out of the water, like flag iris, for creatures such as dragonfly larvae to climb up. You also need creeping plants at the edge to help other creatures to clamber in and out. Don't forget to leave some shoreline bare of plants for birds and insects to collect mud. Then you'll need marginal/bog plants to create a shelter of damp, shady cover.

Try to include *Mentha pulegium*, a creeping mint with mauve flowers, which thrives in mud on the edge of a pond and is a nectar source for butterflies, bees and hoverflies. *Ranunculus flammula* has a sprawling habit and buttercup-like flowers from June to September; it also flourishes at the water's edge. *Callitriche palustris* is a good oxygenator in shaded ponds, forming mats of stems with rosettes of leaves; its cover protects fish and infant frogs. One of my favourites, *Myosotis scorpiodes*, the blue water forget-me-not, grows in boggy soil or shallow water; newts lay eggs on the leaves.

Providing access to the pond via plants, a log or stones is vital for wildlife. Ornamental fish are beautiful, but some wildlife (such as newts) will avoid a pond containing fish. Ideally, allow the pond to populate naturally without introducing 'alien' species.

Will a pond cope with no winter sun?

It is commonly accepted that for an artificial pond to be a thriving, vibrant and healthy environment the amount of sunlight it should receive is about six hours a day in the growing season. This is largely due to the fact that the majority of aquatic habitat plants and water margin plants are sun lovers. The sunlight also warms the water, so increasing the opportunity for plant and animal life within it.

Additionally, ponds shaded by trees can experience problems with fallen leaves accumulating, releasing water-colouring toxins as they decompose, and causing negative adjustments in the oxygen and chemical balance of the water. This is a bigger problem with fish as their waste increases nitrogen levels in the water. If filters and oxygenators are not used the resultant algal soup can choke the pond.

It may be possible to create a successful pond on the basis that the November to March period you describe is a relatively dormant time in the pond. So if, from March to November, the required hours of sunlight are achieved, then success may be possible. If in doubt, site the pond in a sunnier location.

Are there any pond plants I should avoid?

Yes, is the simple answer. Almost any plant can cause a problem – even natives, as anyone who has had their pond overrun with yellow water lilies or bullrushes will tell you. It's obviously a good idea to avoid known offenders, such as blanket and duck-weeds. Other main offenders to steer clear of are the New Zealand pigmyweed, *Crassula helmsii*; parrot's feather, *Myrio-phyllum aquaticum*; floating pennywort, *Hydrocotyle americana*, and the water fern, *Azolla filiculoides*.

Can I use pond blanket weed in the garden?

Some people use it as a lining for their hanging baskets, but it's not nearly as water-retentive as moss. It's more useful to put it on the compost heap or spread it as a mulch on borders. Don't forget to leave blanket weed next to the pond for a few hours after removal to let aquatic creatures escape back into the water.

My pond membrane is leaking – how can I find and repair the hole?

Fill the pond up and let the water level drop until it finds a new level and remains constant – this shows you the line on which the hole is. Run your finger around this line until you find the hole. Use a pond repair kit (available from aquatic centres) to patch the leak. Some people are surprised at the amount of water a pond loses through evaporation, especially when the weather's warm. In a normal summer, a pond's level could need topping up weekly to compensate for this. So make sure this isn't the cause first – make a note of the date, then watch how the water level drops over the course of a few weeks.

Glossary

Acid soil – soil with a pH value below neutral (less than 7).

Aerate – to loosen soil in order to relieve compaction, allowing for improved drainage and movement of air.

Alkaline soil – has a pH above neutral (higher than 7).

Annual – specifically used in gardening to denote a plant that completes its life cycle (germinating, growing, flowering, setting seed and dying) within a single growing season.

Annual weeds – weeds that complete their life cycle within a year, such as chickweed and speedwell.

Aphid – minute plant-feeding insects commonly known as greenfly, blackfly or plant lice.

Aspect – the direction in which something faces (north, south etc.).

Bareroot – plants lifted from the ground and sold unpotted in the winter months.

Bedding plant – any plant that is planted out in a bed, border or container for a seasonal display, most commonly during the summer and spring months.

Biennial – a plant that completes its life cycle over the course of two growing seasons.

Biological pest control – using living organisms to control garden pests such as whitefly or slugs.

Blackfly – see aphid.

Bolt – when a plant goes to seed prematurely.

Botrytis – a fungus that causes disease in a wide array of herbaceous, annual and perennial plants. It is characterised by a grey mould.

Bud – an undeveloped or embryonic shoot that normally occurs in the axil or at the tip of a stem.

Bulb – a short, underground stem with fleshy, modified leaves that is used as a food storage organ by a dormant plant.

Canker – this is a general term for a disease that's characterised by patches of dead cells on either the trunk or branches of a tree or a woody plant.

Chitting – to encourage seed potatoes to sprout before planting, by placing them in a bright spot until the shoots are about 3cm long.

Cloche – a covering that is used to protect plants from cold temperatures, frost, birds and pests.

Club root – a disease affecting cabbages, radishes, turnips and other members of the mustard family (Cruciferae). It leads to undeveloped heads, or a failure to head at all.

Cold frame – a frame covered in glass or plastic that is used to protect plants and seedlings outdoors.

Companion plants – plants that are beneficial for one another, with one plant perhaps repelling pests that prey on the other.

Compost – 1) a shop-bought potting medium to grow plants in; 2) a homemade organic material made from a mix of brown and green waste left to decompose.

Container grown – refers to plants grown and marketed in containers of varying sizes. They may be planted all year round.

Coral spot – a fungus that usually colonises dead stems but can attack live twigs and branches, causing them to die back.

Cordon – a space-saving plant, usually a stone fruit, with its growth restricted to one unbranched stem that may be trained vertically or at an angle.

Corm – an underground storage organ, comprising a short, swollen portion of stem with a protective skin known as a tunic. Corms are planted like bulbs.

Crown – 1) the section of a tree above the main stem, comprising live branches and foliage; 2) the part of a plant where the root and the stem meet.

Cultivar – the name given to any plant variety that has been cultivated to distinguish it from a wild species.

Cutting – also known as a slip, a cutting is taken from a healthy
plant by using secateurs or a knife. The cutting is then placed in
a growing medium to create a new plant.

Damping down – refers to the practice of spraying the floor of
a greenhouse or polytunnel on hot mornings to increase the
humidity inside.

Damping off – the term used for fungal ailments, usually affecting
seedlings by causing the stem to rot off at soil level.

Deadheading – removing dead flower heads to encourage more
blossoms.

Deciduous – used to describe certain plants, principally trees and
shrubs, that shed their leaves seasonally.

Dieback – the term describing the dying of the outer portions
(the tips and branches) of a plant as a result of either disease or
climatic conditions.

Divide/division – the means by which herbaceous stock may be
increased by dividing the clump into a number of sections.

Dormant – when deciduous trees, shrubs and herbaceous perennials
stop growing over winter (usually October to March).

Drill – a straight, narrow trench where seeds are sown.

Earthing up – to 'earth up' is to cover roots with heaped-up soil,
especially potatoes, to prevent greening of the tubers.

Ericaceous – plants such as blueberries that are lime-hating and
require acid soils.

Espalier – usually refers to a fruit tree, with branches trained horizontally and flat against a wall or other framework.

Evergreen – plants with leaves that remain green and functional
all year round.

Fan – a method of training plants against a wall or on wire to achieve
a fan-like effect.

Fertiliser – concentrated plant nutrients used to improve growth,
flowering and fruiting.

Frost pocket – a small, low-lying area, where late and early frosts are more likely, increasing the risk to tender plants.

Genus – a group of closely related species.

Germination – following fertilisation, germination is the sprouting of a seed into a seedling.

Graft – the insertion of a section of one plant, usually a shoot, into another so that they can grow together into a single plant.

Greenfly – see aphid.

Ground cover – low-growing, fast-spreading plants, often used to stabilise soil and prevent soil erosion as well as to provide aesthetic interest in an area of garden.

Half-hardy – a plant that will tolerate cooler temperatures. Half-hardy plants may be grown outdoors in summer but would not survive frosts, while half-hardy shrubs and herbaceous plants may survive an average winter if in sheltered or climatically favoured environments.

Hardening off – slowly acclimatising plants raised indoors to the conditions outside before planting out.

Hardy – a plant that is able to withstand low temperatures without any protection.

Herbaceous – a non-woody plant whose upper parts die back to the soil surface at the end of the growing season each year.

Hybrid – the term used for a plant resulting from the cross-fertilisation of two or more plant species or genera.

Joint – another term for node, the point on a stem from which a leaf or leaf bud grows.

Larva – refers to the post-embryonic stage in the development of an insect before its metamorphosis into an adult; e.g. a caterpillar, grub or maggot. The larval stage is generally considered the most destructive period in the insect life cycle.

Lateral – a bud rising from the leaf axil at a node in the stem which will go on to develop into a side shoot.

Leader – 1) the tip of the main stem of a plant; 2) the main stem from which laterals (see above) rise.

Leaf curl – a fungal disease of peach, almond and nectarine trees, which causes an increase in cell growth, resulting in thick, puckered and curled leaves that change colour from green to yellow and dark red.

Leaf miner – a grub, the larva of small flies and some moths, which tunnels into the leaves of many plants.

Leaf mould – composted fallen leaves or the detritus of partially decayed leaves found under trees.

Leaf spot – a symptom of a variety of viral diseases that attack the leaves of many plants, particularly roses. The spots can be various colours.

Liming – adding lime to soil in order to improve its structure, reduce acidity and remedy a calcium deficiency.

Loam – a rich, fertile soil with a balance of sand, clay and humus.

Loam-based composts from the John Innes range are soil-based, rather than peat-based. They have good moisture and retain nutrients well.

Mealybug – an insect resembling a wood louse but which is coated in a pale, waxy substance. Feeds on sap.

Mildew – a fungal plant disease that causes a powdery coating on the surface of affected plants. Commonly occurs when plants have been exposed to damp conditions. It can be prevented by good ventilation.

Mulch – material added to the surface of the soil to lock in moisture and suppress weeds.

Nectar – the honey-like secretions of plants, commonly found in flowers, that attract bees and insect pollinators.

Neutral – pH7; a soil that is neither acid nor alkaline.

Node – the point on a stem from which the leaf or leaves rise.

Organic – refers to the matter derived from the decay of vegetation, animal waste and animal tissue.

Organic gardening – the cultivation of plants without the use of chemicals of inorganic origin.

Oxygenator – an aquatic plant that releases oxygen into the water when it is submerged.

Peat-free composts commonly consist of composted bark, wood fibre and coir with added nutrients.

Perennial – usually refers to a non-woody plant that lives for more than two years or three seasons. Flowering annually, perennial plants tend to die down during the winter, but have roots that will survive low temperatures.

Perlite – a lightweight aggregate made from an amorphous volcanic or other vitreous rock. Totally inert and sterile, it is useful in opening cavities in soil and enabling water and air to reach plant roots. Often used in potting soil.

pH – this symbol is followed by a number to indicate the degree of acidity or alkalinity in soils or solutions. Related to a standard solution of potassium hydrogen phthalate, pH7 is neutral. Numbers below 7 indicate acidity, while higher numbers suggest alkalinity.

Pinching out – method of stopping a plant from growing upwards by removing the growing tip, thereby encouraging the development of side shoots.

Plant up – to transfer a plant to a new pot.

Plug plant – a small but well-rooted seedling raised in a cellular tray; useful for planting large areas.

Pollination – the transfer of pollen (usually by insects, but also by wind or hand) to the female stigma, after which fertilisation of the ovule ensues.

Pot bound – a plant that has outgrown its pot – its roots are restricted and growth will be become stunted.

Potting on – the transfer of a plant from one pot to a larger one (usually the next size up) to allow more space for the plant's roots to develop.

Pricking out – transplanting seedlings from the seed beds or pots in which they were sown to new, larger receptacles. The term is derived from the old practice of pricking small holes in the soil in preparation for the planting of young seedlings.

Propagation – the processes by which plants may be increased in number, including grafting, division, cuttings, seeding, budding and air layering.

Red spider mite – tiny red spiders, sap-sucking pests that are a problem in greenhouses and on house plants.

Rhizome – a creeping stem that grows along, or just beneath, the soil surface, with roots rising from it. A rhizome also serves as a storage organ.

Root fly – an insect that attacks plant roots. May be controlled by use of pesticides.

Root nodule – knobbly growth rising from the roots of leguminous plants. Root nodules contain bacteria, used in nitrogen fixation.

Rootball – the matted roots and soil of a plant, which should be kept in its entirety when transplanting a pot-grown plant.

Rootstock – the plant onto which a scion from another specimen may be grafted; the rootstock thereafter provides the root system of the new combined plant.

Rots – general term for a number of diseases that cause the death or decay of plant tissue.

Rust – a fungal disease of plants, causing rust-coloured spots on leaves, stems and other parts.

Sawfly – carnivorous flies, many of which feed on nectar.

Scab – a fungal scale disease, causing lesions on leaves or fruit.

Scarify – raking to remove thatch and moss from your lawn.

Seed – the fertilised and matured ovule of a plant containing an embryo and sufficient nutrients for its development.

Seedling – a young plant; one that is cultivated from seed (as opposed to one derived from a cutting).

Self-fertile – a plant that has the ability to pollinate its own flowers.

Shrub – a multi-stemmed woody plant with several branches or shoots rising from the base but no single trunk.

Silver leaf – a disease primarily affecting plum trees.

Species – a group of plants with common characteristics.

Specimen (plant) – refers to a plant of superior quality which is put in a prime position to show it at its best.

Spore – a simple reproductive cell of non-flowering plants, e.g. ferns, mosses and fungi. Has the function that's correspondent to the seed in higher plants.

Spur – 1) a long, tubular projection at the base of a petal that may contain nectar; 2) a large lateral root, or the branch of a root.

Staking – the practice of supporting young or fragile plants by loosely attaching them to canes or stakes driven into the ground.

Stopping – method of pruning, whereby the plant's tip is removed to encourage the production of side shoots.

Tap root – the large, central root that grows downwards and from which smaller, lateral roots grow.

Tender – describes plants that are unable to stand frost or freezing temperatures.

Tendril – the coiled, cord-like growths (modified stems or leaves) produced by climbing plants, which enable them to attach themselves to a support or trellis.

Thatch – a layer of dead moss, old grass stems and other debris in your lawn.

Thinning out – reducing the number of plants in a bed or container to provide more room for growth. In fruit production, it means to reduce the crop of fruitlets early on to produce larger fruits for harvest.

Top dressing – improving soil by adding a layer of fertiliser to the surface and allowing it to settle in without digging over; also means to replace the top layer of soil with compost.

Topsoil – the strata of soil nearest the surface and down to a depth of around 30cm.

Transplanting – moving a plant from pot to pot, or into the garden.

Trench – a long, narrow ditch dug out of the ground.

Trunk – the woody main stem of a tree.

Tuber – an underground storage organ used by some plants to provide energy and nutrients for regrowth the following season.

Vermiculite – a mineral that forms spongy, lightweight kernels that are useful in preventing the compaction of soil and in loosening heavy soil. It is also used as a medium in which to root cuttings. Vermiculite enables soil to hold more water.

Vine – a climbing plant; a grape vine.

Virus – a group of simple, microscopic organisms that reproduce inside the cells of plants, thereby destroying them and causing disease. Viruses that attack plants are usually transmitted by sap-sucking insects.

Weed – any unwanted plant; one that hinders the growth of more desirable plants.

Weevil – a member of the Curculionidae beetle family. Small in size with an elongated head, weevils can cause lots of damage to fruit, nuts, trees etc.

Whitefly – a sap-sucking insect. Whitefly are particular pests in the greenhouse as they can be destructive.

Index

182, 194, 235, 237–8

fruit 179–95
  bare-root 185–6; feeding 187–9;
  low/no yields 182, 188, 192,
  219; pollination 182, 187, 189,
  192–3; pruning 187–8, 192, 219,
  225–6; thinning out 192–3, 194;
  unripe 188, 193–4; *see also specific*
  fruit

fruit cages 96, 191
fruit trees 111, 185, 190–1, 194, 225
  see also specific types of fruit tree
fuchsia 92, 226, 235
fungal diseases 109–19, 146, 171,
  173–6, 184, 206–7, 222, 225, 227,
  235, 241–3
fungi 16–17, 19, 26, 36, 219
  mycelium 16; mycorrhizal 46,
  112

garden design 121–40
  themes 132
garlic 205, 210
geranium 18, 51, 138–9, 150, 169
  see also pelargonium
gooseberry 191, 194, 195
grasses, ornamental 128, 132–4,
  139, 226
green roofs 128
'green walls' 130–1
greenhouses 208, 212, 231–45
  heating 234–5; humidity 208,
  234–7, 242; hygiene 112, 117,
  235, 240–1; mini 244–5; shading
  234, 238, 243; temperatures
  234–5, 244; ventilation 233–4
ground cover 77, 259–60
ground elder 74–6, 80, 84

Hamilton, Geoff viii-ix, x
Hamilton, Nick ix
hanging baskets 40, 47–9
hedges 125, 128–9, 226
herbs 159, 245
hoeing 78, 82, 83–4
holidays 165–6
honey fungus 109, 114, 115
honeydew 236

hornwort 257, 259
horsetail 75, 80, 85
horticultural grit 157
hosta 156, 158
houseplants 55–69, 92
  brown leaves 58–9; dead 59, 69;
  and dust 62, 66; feeding 60, 65;
  health benefits 57, 68–9; and
  humidity 61; and lighting 65,
  66; and pests 59–60; poisonous
  68; repotting 62, 63; watering
  58–61, 64
hoverflies 91, 93, 205, 260
humidity 59, 61, 118, 176, 183, 208,
  234–7, 242
hydrangea 110, 118–19, 136, 218,
  228–9

iris 138, 146, 257
ivy 80, 130, 205

jasmine 124, 226

Klein, Carol 49, 156, 201, 211, 253

lacewings 91, 102, 205
ladybirds 102, 104, 204
lavender 51, 99, 103, 152, 226
lawns 1–20
  aeration 1, 4, 9, 17; bare
  patches 5–6; brown 7, 16, 17;
  compaction 1, 4, 13; drainage
  9, 17; edging/re-shaping 6, 11;
  fairy rings 16–17, **16**; feeding
  8–9, 17; leaving grass clippings
  on 15–16; level 5, 19; and moss
  5, 9, 13; mowing tips 6–8, 12–15;
  and pests 98, 105; reducing the
  size of 18; scarifying 5, 9, 10,
  17; sowing new 19–20; thatch
  5, 14, 19; top-dressing 9–10, 19;
  watering 7; and weeds 4, 8, 9, 17,
  82; wildlife-friendly 17–18
leaf curl 109
leaf drop 58
leaf scorch 58
leaf spot 118, 222
leafmould 27–9, 155